Hiroshi Hara
The 'Floating World' of his Architecture

Hiroshi Hara

The 'Floating World' of his Architecture

Botond Bognar

WILEY-ACADEMY

All project descriptions are by Hiroshi Hara

English translations of Hiroshi Hara's essays and project descriptions appearing here on the pages indicated below are by:

Kazuhito Furuya: pp 34-37; 55; 56-59; 86-89; 120-123; 162-165; 202-204; 242-244; Kazuhito Furuya and Sheri Blake: pp 146; 150; Hiroshi Watanabe: pp 38; 42; 46; 50; 60; 66; 78; 90; 102; 108; 124; 132; 152-153; 172; 186; 206; 216; 226; 230; 246; 250; Koichiro Ishiguro: p 254; Takashi Yanai and David B. Stewart: pp 82; 166.

Previously published texts by Hiroshi Hara have been reprinted here on the pages indicated below by courtesy of the following companies which are also the Copyright holders of these texts:

Shinkenchiku-sha Co Ltd Tokyo: p 38 (originally in *The Japan Architect* Nov. 1972).

A.D. A. EDITA Tokyo Co Ltd Tokyo: p 82 (originally in *GA* Architect 13 Hiroshi Hara); p 128 (originally '500m x 500m x 500m - 1992' in *GA* Japan 02); p 166 (originally 'Media Park Köln' in *GA* Architect 13 Hiroshi Hara); p 168 (originally 'La Cité Internationale de Montreal' in *GA* Document 29); p 170 (originally 'Entry in the BIT Competition' in *GA* Document 43).

The Shokokusha Publishing Co Ltd Tokyo: p 132 (originally 'Homogeneous Space as Column' in *Kenchiku Bunka* Special Issue on Mid-Air Garden - Interconnected Skyscraper 1993: Document of Shin Umeda City).

Kajima Institute Publishing Co Ltd Tokyo: p 146 (originally 'Three Significant Aspects of Extra-Terrestrial Architecture' in *SD Space Design* 01/94, Special Feature on Hiroshi Hara: Potentia of Architecture).

City of Sapporo, Japan p 254 (originally in a Pamphlet).

Photographic credits

Every effort has been made to locate sources and credit material but in the very few cases where this has not been possible our apologies are extended: Atelier Phi pps 7, 71, 206, Botond Bognar pps 10-33, Mitsumasa Fujitsuka pps 139, 142-143, GA Photographers pps 226, 250-251 (bottom), Hiroshi Hara pps 9, 225, 255, 256-257, 263 (bottom right), Manoru Ishiguro pps 186 (top), Kazuhiro Kojima pps 247, 248-249, Manoru Mishima pg 175, Ryuji Miyamoto pps 140, 141, Takashi Miyamoto pps 43, 50 (middle), 55 (bottom right & left), Osamu Murai pps 124, 125, ©NASA pps 121, 147, 147 (inset bottom right), 149 (left), 150-151, Tomio Ohashi pps cover jacket, 2, 47, 48, 50 (top), 55 (top), 60-65, 67-70, 72-83, 87, 89-119, 133-138, 144-145, 152, 153 (left), 157 (top), 158-161, 167-169, 174, 176-177, 182-185, 187, 190-197, 200, 207-215, 217-219, 220-221 (top), 224, 254, 262, 263 (top right), 264-271, Shigeru Ono pps 128, 130, 147 (inset bottom left & centre), 148 (bottom), 170, 222-223, Asashi Shimbun pg 8, Shinkenchiku pps 40-41, 153 (right), 154-156, 157 (bottom), 172-173, 180-181, 186 (bottom), 188-189, 201, 216, 220-221 (bottom), 227-229, Shoukokusha pps 39, 129, 230-241, Syuji Yamada pps 6, 42, 44-46, 49, 51-54.

Cover: Sapporo Dome, General view from the north
Page 2: Umeda Sky Building, Looking up towards the sky garden

First published in Great Britain in 2001 by
WILEY-ACADEMY

a division of
JOHN WILEY & SONS LTD
Baffins Lane
Chichester
West Sussex PO19 1UD

ISBN 0-471-87730-1

Copyright © 2001 John Wiley & Sons Ltd. All rights reserved.

No part of this publication may be reproduced, stored in a retrieval system, or
transmitted, in any form or by any means, electronic, mechanical, photocopying,
recording, scanning or otherwise, except under the terms of the copyright, Designs and
Patents Act from 1988 or under the terms of a licence issued by the Copyright
Licensing Agency, 90 Tottenham Court Road, London UK, W1P 9HE, without the
permission in writing to the publisher.

Other Wiley Editorial Offices
New York • Weinheim • Brisbane • Singapore • Toronto

Design and Prepress: ARTMEDIA PRESS Ltd, London

Printed and bound in Italy

CONTENTS

10 **Between Reality and Fiction**
The 'Floating World' of Hiroshi Hara's Architecture

34 **1. Reflection / Embedding**
38　Awazu House
42　Hara House
46　Kudo Villa
50　Niramu House
55　Villa *Yume-butai*, or the Stage of Dreams

56 **2. Multilayered Space / Self-Similarity**
60　Josei Primary School
66　Yamato International
78　Yukian Teahouse
82　Future in Furniture

86 **3. Transposition / Incidental Illusion**
90　Tasaki Museum of Art
102　Kenju Park 'Forest House'
108　Iida City Museum

120 **4. Floating / Mid-air City**
124　Ito House
128　500 x 500 x 500-Metre Cube
132　Umeda Sky Building
146　Extra-Terrestrial Architecture
152　Four Cube-Houses

162 **5. Attractors / Semiotic Field**
166　MediaPark Köln
168　The International City, Montreal
170　Parc BIT, Majorca
172　Miyagi Prefectural Library
186　Kyoto Station Complex

202 **6. Traversing / Bridging Movement**
206　Ose Middle School
216　Motomachi High School
226　Takagi Clinic
230　University of Tokyo Komaba II Campus

242 **7. Modality / Possibility of Events**
246　Modal Space of Consciousness/Robot Silhouette
250　Twenty-five Music Stands
254　Sapporo Dome

271　List of Works
272　Biography and List of Collaborators

Between Reality

and Fiction

The 'Floating World' of

Hiroshi Hara's Architecture

Hiroshi Hara

Iida City Museum, Iida, Nagano Prefecture, 1988. Rooftop detail

BETWEEN REALITY AND FICTION
The 'Floating World' of Hiroshi Hara's Architecture

Botond Bognar

Hiroshi Hara explores a floating world which is no longer based on physical or even cognitive reality, but would seem to evoke the erratic images of our deepest mental structure. He takes us ever more convincingly beyond the realm of the stable and permanent, and into the volatile boundaries of human consciousness, where cryptic and ephemeral images give a shock to solid reality. Shifting from one mode of vision to another, from real to unreal, his buildings express a state of dreamlike ferment amid tangible things, and the hovering light invokes here, as it did in the Japanese past, what Hara calls aptly a 'fusion of reality and fiction'.[1]

It is only through a change of human consciousness that the world will be transformed.[2]

INTRODUCTION / BACKGROUND

In the past few decades, Japan has nurtured an increasingly large number of unusually talented and innovative architects. Attracting significant international attention, contemporary Japanese architecture has achieved a leading position in the world of design and construction. This was especially evident during the 1980s and early 1990s, a time that saw an extraordinary quantity and broad range of quality in built structures. Riding on the boom of the 'bubble economy', architects enjoyed the seemingly limitless patronage of eager clients with deep pockets, and were free to engage in various design experiments. This runaway architectural culture and period of accelerated urbanism have yielded, not surprisingly, both hordes of inferior, frivolous buildings, and, more significantly, innumerable exceptional projects that have justifiably earned international recognition and admiration. In addition to the already familiar architectures and commanding reputations of Kenzo Tange, Kazuo Shinohara, Arata Isozaki and Kisho Kurokawa, the stunningly powerful individual works of such designers as Fumihiko Maki, Tadao Ando, Toyo Ito, Itsuko Hasegawa, Shin Takamatsu, Riken Yamamoto, Kazuyo Sejima and many others have become widely known and highly influential within and beyond the confines of Japan (figs. 1–5). This spectacular 'new golden age' of Japanese architecture has also witnessed the maturation and flourishing of Hiroshi Hara's architecture, which has constituted one of the most original voices on the increasingly variegated stage of Japanese design since the 1960s.

Hara has made a considerable contribution to shaping the course of Japanese architecture. Equally, the new and rapid developments in the social, cultural and urban conditions of the country have influenced his work, prompting a gradual shift in direction and broadening in scope of his design. Starting primarily with small residential buildings in the 1960s, today he is responsible for the design of some of the most significant large-scale Japanese projects of recent years. Ranging from a small teahouse to extensive urban complexes, his oeuvre includes art galleries and museums, schools and university campuses, clinics and hospitals, highrise and other multipurpose offices and commercial buildings, a huge station complex in Kyoto, a major library in Sendai, and a soon-to-be-completed sports and entertainment facility in Sapporo. Hara has received many awards and prizes such as the Annual Award of the Architectural Institute of Japan, and the first Togo Murano Award. Moreover, he has participated in and won innumerable major international competitions, and has exhibited his work and lectured all over the world. Equally important is the fact that he has been involved in substantial theoretical investigations, which he launched while teaching as a professor in the University of Tokyo, and continued until his retirement from academia in 1997. Combining academic research with his fully fledged design practice has had a profound impact on the type and quality of work he has produced.

Hara's designs, inspired by many sources and shaped by his acute sensitivity to the world around him, explore a multitude of issues. He has addressed these both in response to changes in external conditions and in relation to the stages in his own development as a thinker and designer. Consequently, his works

1. Fumihiko Maki: Fujisawa Municipal Gymnasium, Fujisawa, 1984. Exterior

2. Tadao Ando: Water Temple, Awaji Island, Hyogo Prefecture, 1991

reveal, on the one hand, parallels with those of some of his contemporaries, reflecting current issues in architectural discourse, and most significantly, on the other hand, features that are unique to his own architecture. This is particularly true with regard to the experiential qualities of space and place. It is important to recognise Hara's long-standing scholarship and unremitting commitment to developing various design strategies in the field of architectural phenomenology in order to 'map' the faculties and workings of human perception or consciousness. Throughout the years, he has maintained a strong interest in probing the boundaries and/or the relationship between the real and the fictive in architecture to search for the possible meanings in the environment, natural and built, while aiming at re-establishing their unity.

Born in 1936, Hara entered the stage of Japanese architecture in the 1960s as one of the representatives of a new generation of architects, often referred to as the New Wave, who, putting forward a plurality of unorthodox approaches to design, began to play an increasingly significant role in the 1970s. These were times of considerable turmoil and ferment in Japan; as a result of the oil shock and the ensuing energy crisis, the economy suffered, just as today, from an insistent recession. Simultaneously, rapid and unchecked industrialisation and its technologies, along with the explosive urbanisation that had been mushrooming since the late 1950s – products of the 'economic miracle' – betrayed their destructive elements, particularly with regard to the natural and built environment. All these events rapidly began to erode a previously unshakeable faith in the redeeming capacity of the all-encompassing project of modernity, triggering significant changes in social awareness, intellectual outlook and cultural developments in the country. Grounded in the progress of industrial technology, the ideological basis of the Metabolist movement – a late representative of modern architecture in Japan – was already bankrupt. The heroic age of the 1960s was over. In what signalled the advent of a new era, architects and urbanists were prompted to search for new sources of inspiration and concrete solutions for the crisis in architecture and the city.

Such developments were not limited to Japan; they emerged with equal force right across the industrialised world. Countries belonging to the 'first world order' experienced most strongly the passing of the industrial age, the eclipse of the modernist project, and a profound breakdown of modern architecture, especially its urban-design principles. Modernism gave way to Postmodernism and a plurality of design approaches, many of which exhausted themselves with superficial solutions, treating the symptom, rather than the cause of the illness itself. Yet, other alternatives were also explored. The old guard in the Congress of International Modern Architecture (CIAM) was succeeded in the 1960s by Team Ten, a group of reformists led by the prominent Dutch architect and theoretician Aldo van Eyck. This is important to note, because, along with his consistent and significant criticism of modernist dogmas, he was the first to turn attention to the vernacular as a viable source from which contemporary architecture could renew itself. Van Eyck's profound interest in architectural anthropology led him to initiate his explorations of the indigenous architecture of Dogon villages and other types of settlements in Mali, with his first trip to Africa in 1958, followed by others later on. In so doing, he almost single-handedly launched a movement in architecture that was significantly influenced by new developments in anthropology.[3]

As Japan recovered from the devastations of the Second World War, Japanese architects such as the quintessential modernists Kenzo Tange and Kunio Maekawa became more and more active, first at home, then on the international scene, participating initially in several major CIAM conferences abroad. Others, representing a younger generation, including such prominent designers of today as Fumihiko Maki, Minoru Takeyama and Yoshio Taniguchi, continued their studies in the United States, where many of the contemporary architectural developments were unfolding.[4] Subsequently, some of them, including Maki and Takeyama, also gained work experience abroad before setting up their practices in Japan.

YUKOTAI / THE BEGINNING

Although some eight years younger than Maki, and despite the differences in their architectures, Hara started his career with very similar aspirations and intentions. Both Maki and Hara graduated from the University of Tokyo, where they also taught for many years, though in different departments.[5] Both became actively involved in extensive theoretical investigations, related particularly to the pressing issues in urban design – perhaps even more urgent in Japan than in other industrialised countries – resulting from the failure of modernist practice. Significantly, Hara, just like Maki before him, found an exit from the dead-end of Modernism in the lessons taught by the vernacular architectures of innumerable villages in the varied regions and cultures of the world. This is what he would principally investigate with his students almost until the end of his teaching career. The first of his many trips in 1972 led to various Mediterranean countries, followed by repeated visits to South America, the Middle East, several Arab countries, Indonesia, Papua New Guinea, East Africa and China, in addition

4. Riken Yamamoto: Rotunda, Yokohama, 1987

3. Toyo Ito: Shimosuwa Municipal Museum, Shimosuwa, 1993

5. Kazuyo Sejima: Saishunkan Seiyaku Women's Dormitory, Kumamoto, 1991. Interior

to many parts of Japan. He returned to the Mediterranean region several times.[6] Commenting on this work, Hara wrote in 1999:

> We tend to think that architecture history is primarily about temples, shrines, palaces or public building. However, villages are equally beautiful and sparkling. Moreover, villages teach us lessons about how the cities we live in could be designed or even about how we should live on this earth in the context of the environmental crisis ...[7]

Before reviewing the results of Hara's research into the architecture of villages, we should first discuss his early theoretical writings, dating from before his trips abroad. He began to develop his *yukotai* theory in 1965, subsequently refining it over a long period. *Yukotai* refutes the validity of Modernism's homogeneous space and its unity of form. Hara's proposal gives a more important role to the basic constituent elements of architecture – in which the individual spaces are considered as 'closed domains'. The overall composition is allowed to unfold according to 'operational factors: moving elements or forces' and a system of 'apertures' that controls those factors.[8] When the design proceeds from the parts, a preconceived final form – characteristic of orthodox modern architecture – cannot control the whole. As an 'incongruous' aggregate of its 'autonomous' parts, it is the outcome not only of the nature of the elements themselves, but also, and very importantly, of their variable relationships. In Hara's model, these connections are provided by a 'floating domain', an in-between space comprised of largely circumstantial events and their indefinable movements, which can be likened to Brownian movement. It is therefore heterogeneous in nature and 'amorphous' in form; it makes up an interconnected yet incidental system of gaps or cavities in the body of the unfinished whole. The term *yukotai*, meaning 'porous body', refers to this quality. Of his conceptual project City of Holes he wrote: 'The basic nature of architecture is in its holes. The geometric relations between open and closed determine what the piece stands for.'[9]

This model of organisational and/or spatial quality was what Hara and his team were looking for in the patterns of indigenous villages and towns, which they set out to investigate in great detail. David Stewart has pointed out that 'At times ... it did seem as if what the [Hara] group were after was something of the order of the "pattern language" that has been meticulously sought and developed for so many years by Christopher Alexander and his associates', and which carried the more scientific methods of Structuralism.[10] It is true that Hara's work did initially betray traces of a similar interpretation to Alexander's, but, as Stewart points out, when faced with the reality of concrete examples, Hara re-examined and altered his strategy. As opposed to Alexander's data-gathering by way of 'a precise mathematical description of the context ... instead of trying to make a diagram of forces whose field we do not understand', Hara gradually opted for a more complex topological and phenomenological mode of investigation.[11]

In the process, Hara learned many lessons, as he himself has acknowledged.[12] His architecture has been from the very beginning influenced significantly by many attributes of the vernacular in the world. Interestingly, no specific indigenous culture is identifiable in his works; rather, he seems to distil and filter all the many features into a language that ultimately re-presents them in a substantially new way. This particular interpretation of vernaculars, in which references are unmistakable but not clearly identifiable, is seen not only in Hara's early designs for small residences, but later on also in his larger, often urban-scale projects. However, by way of the relationship between elements created by fluid or 'floating' domains, and of a manifest lack of interest in creating dominant forms, the design process is guided largely by topological considerations.

Hara's first three completed projects, the tiny Ito House of 1967, the Shimoshizu Primary School and the Keisho Kindergarten of the following year, may have been very early attempts to implement the *yukotai* theory, yet the outcome in each case was nothing short of a radical architecture. As early as 1968, the critic Robin Boyd found these works to be 'disturbing, challenging, promising, but as yet not mature enough to suggest a new direction'.[13] Boyd is certainly right in most of his observations, but he misses the already unmistakable 'new direction' manifest in these works, namely a sharply anti-Metabolist stance; a debunking, as in other works of the New Wave, of this prevailing and strongly (industrial) technology-oriented movement that ruled Japanese architecture in the 1960s.

While constructed of different materials, these buildings have been shaped with an emphasis on easily recognisable constituent spatial units and their amorphous, free-flowing connecting and enveloping spaces. In all three projects, the special articulation and positioning of openings – 'apertures' in Hara's terminology – skylights, windows and doorways, play important roles in defining the individual elements. These refuse to add up to a unified or synthesised form, instead yielding truly incongruous and fragmentary 'compositions'. Chris Fawcett, writing about these early buildings, recognised that the school buildings are

> characterised by a constant interpenetration of parts by the whole and their perpetual dissociation and adaptability, the whole sequence rendered comprehensible by a linear processional runway that threads the incidents together; in other words, at the foci of this dynamic flux are points of harbouring and gathering. We can sense, for example, in the Keisho Kindergarten, that Hara envisages architecture as a link in a chain of concrete action.[14] (figs. 6-8)

Here form is obviously not composed with any preconceived ideal in mind; rather, it is the chance outcome of 'induction', a method in which each part is situated according to its dynamic role, with a certain disregard for the generated whole. The process of generation was further underscored in the case of the Ito House, where an extension had to be built later, but which addition did not – *could* not – alter the already open form.

The innovative feature of Hara's new mode of design – as well as that of Maki, Van Eyck and others – is that by rejecting the compositional principles or, in effect, the metaphysics of modern architecture and urban design, and researching vernacular architecture and architectural anthropology instead, these architects opted for a pre-industrial, traditional model for their urban architecture. Maki and Hara, however, differ in terms of the languages they apply in their designs. From the very beginning of his career, Maki has relied on the vocabulary of modern architecture, while Hara has, for the most part, adopted a more variegated or hybrid language within the postmodern paradigm. This preference prevailed until very recently, when, in his latest projects, particularly in the Cube Houses (1998), he makes novel use of a sensitive, but more minimalist modern language. This is not to say, of course, that Maki's consistently modern language is poorly articulated and limited, or that Hara's is not a highly refined or sophisticated system; both have much enriched the language of recent architecture in their own particular ways, exploring with great success the expressive possibilities of the latest information and media technologies.

It is perhaps Hara's Josei Primary School in Naha, Okinawa (1987), that comes closest to the earlier, more regulated designs of Van Eyck and Maki. Nevertheless, significant differences remain. For example is Hara's application of elements belonging

8. Shimoshizu Primary School, Sakura, Chiba Prefecture, 1968. Exterior

6-7. Keisho Kindergarten, Machida, 1968. Exterior and first-floor plan

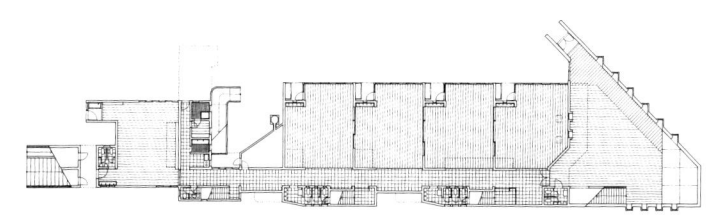

to the vernacular architecture of this tropical island – seen in the red-tiled and decorated roofs over the classroom modules – but again, not without subjecting these sources to some important transformations (fig. 9).

Returning once more to Hara's first works, it can be observed that the spatial formation of the Ito House, and especially the curving, prismatic space of the common living area, has been designed so as to allow for the best airflow, as demonstrated by his experiments using a small-scale model of the house (fig. 10). But Hara also employs the flow of air as a metaphor for the movement of people and events, which cannot be precisely decided in advance. In other words, the issue of immeasurability or *undecidability*, begins to figure even in Hara's earliest work.

The Ito House and Hara's first two educational buildings display numerous other ideas and features that were to be further developed in subsequent works. The residence is a precursor of the tiny houses that he would turn out in growing numbers in the 1970s. If for no other reason than its natural wood construction with freestanding internal pillars and special detailing, this house immediately reminds the observer of its origins in 'primitive' vernacular architecture. Outside, it has an ordinary clapboard finish; inside, plywood panelling is fixed in place with an unusual system of thin battens. Both are suggestive of indigenous structures, yet as soon as one identifies them as such, one recognises that every possible detail is, more or less, peculiar or 'distorted', and a strange feeling of uncertainty seeps into one's awareness. Moreover, the direction of the battens, along with the way in which the wooden floorboards are laid, indicates and enhances the flowing character of the space (fig. 11).

One should also mention the strange roof, designed according to a complex folded and convex system of units in the form of polyhedrons composed entirely of triangular planes. As the surface of the roof – regarded by Hara as a unique *topos* – becomes increasingly significant in his later works, such fragmented, or otherwise amorphous roof structures emerge time and again, even in major non-residential projects. Similar roof forms are seen in the Tasaki Museum of Art (1986), the Yamato International (1987), Iida City Museum (1988), Miyagi Prefectural Library (1997), the Sapporo Dome (2001), and many others, to be discussed later. Indeed, the roof is assigned multiple roles in Hara's architecture (fig. 12).

9. Josei Primary School, Naha, Okinawa, 1987. View of roofscape

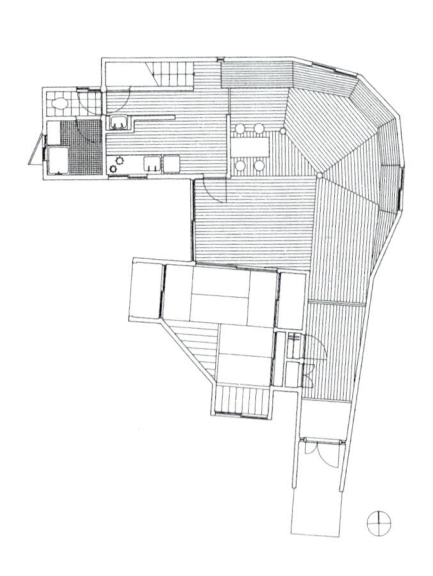

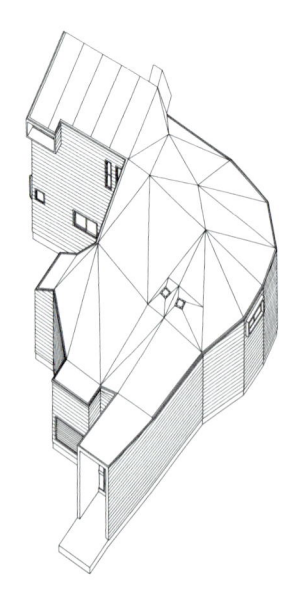

11–12. Ito House, Mitaka, Tokyo, 1967. First floor plan and Axonometric drawing

10. Ito House, Mitaka, Tokyo, 1967. The world of yukotai

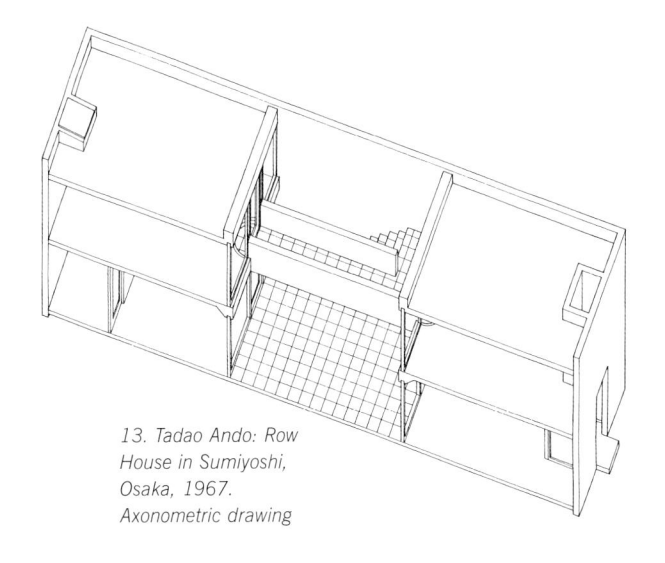

Between Reality and Fiction

13. Tadao Ando: Row House in Sumiyoshi, Osaka, 1967. Axonometric drawing

14. Kiko Mozuna: Mirror Image House, Niiza, Tokyo, 1980. Interior

15. Kazuo Shinohara: House in Uehara, Tokyo, 1976. Exterior

16. Toyo Ito: White U House in Nakano, Tokyo, 1976. Exterior

INTERNAL DREAMSCAPES / REFLECTION HOUSES

The 1970s introduced a new chapter in Hara's architecture. During this time, he worked almost exclusively on small residences. Featuring an exhibition space and related facilities, the only exceptions are the small Shokyodo (1979) and Sueda Art Galleries (1981); yet even these form extensions to houses. The house designs are as much a representation of the next step in Hara's unfolding architecture, increasingly guided by the tenets of phenomenology, as they are his response to developments taking place within the deteriorating urban conditions of Japan. In light of such a hostile state of affairs, an era of protest and 'protection' arrived. Numerous New Wave architects began to shape their

This feature was demonstrated in a rudimentary way by the Keisho Kindergarten. In addition to the several unusually shaped pyramidal or pitched roofs with skylights, there are interwoven flat surfaces on the roof that can be utilised as extensions of the playground below – the ultimate space of flows. The roof terraces and the playground are connected by, and directly accessible through, a wide outdoor stairway. The shaping of the rooftop as a surface *(topos)* continuous or analogous with the land or the city, returns in more pronounced forms in projects like the Yamato International, Iida City Museum, not to mention such obvious examples as the Umeda Sky Building (1993) and the Kyoto Station Complex (1997).

Last but not least, the design of the Ito House features an uncommon spatial element: the second-floor bedroom over the kitchen juts out and connects to the living room through an internal window, alluding to the notion of a space within a space. Such articulation of interior relationships becomes more dominant and is pushed to the limits of its architectural possibilities in Hara's work during the next ten years.

17. Tadao Ando: Koshino Residence, Ashiya, Hyogo Prefecture, 1982. Interior

18. Takamitsu Azuma: Azuma House, Tokyo, 1967. Exterior

buildings as protective realms. Ando, Shinohara, Hasegawa and many others were expressing their uncompromising rejection of the city and the fast-evolving megalopolitan project, along with its increasingly commercialised and trivial manifestations; they were determined to defend inhabitants and users from the overwhelming intrusions of the outside world and even society at large. Ando's Row House in Sumiyoshi, Osaka (1976) is paradigmatic in this regard (fig. 13).

This trend spread fastest in residential architecture, for two reasons. Firstly, it was the individual resident or family whose privacy was most besieged. Secondly, the younger not-yet established architects received commissions almost solely for small private houses to be built for relatives, friends and others in a relatively close circle. A growing number of these residential buildings came to be designed as introverted structures, or small monastic enclaves, showing little desire to communicate with their frequently unfriendly surroundings. Yet, the New Wave was also part of a broader movement, a strong counter-culture that rejected the ideologies and practices of the Metabolist movement in Japan, and an increasingly dogmatic and hollow Modernism in general. Rejection of Modernism in architecture was linked to a profound disillusionment with industrial technology and its dominance in designing and structuring the built environment. The new architecture was both anti-urban and largely anti-technology. Architects found new meanings in idiosyncratic expressions and private languages. With the New Wave, a refreshingly broad and radically interpreted pluralism entered the Japanese urban stage, and the private residence assumed the role of testing ground for the most avant-garde ideas in design. They ranged from the absurd to the enlightened; between the populist and the austere or minimalist; the highly rational and the uniquely organic, even anthropomorphic; and, of course, there were also those who embarked on a return to historicism and the traditional, either literally in form, or in spirit (figs. 14–16).

Intended as spiritual retreats and places of regeneration, many of these buildings, while shutting out the city, sought intimate contact with nature, or what was left of it. This, however, turned out to be no easy task, demanding great creative ingenuity. Within the hard, defensive shell of usually simple geometric forms, we find various attempts to provide possibilities for the penetration of natural phenomena to activate the 'hermetic microcosm' inside. Hara's Reflection Houses represent a series of profoundly evocative alternatives to other, often equally provocative and/or poetic resolutions, such as those put forward by Ando, Shinohara, Takamitsu Azuma, Kiko Mozuna, Toyo Ito and Hiromi Fujii, to mention but a few (figs. 17–19). Although the principles of *yukotai* that Hara had introduced in the first three projects greatly informed the designs of these buildings, these ideas, combined with his new phenomenological operations, found their site primarily inside the realm of the house; in other words, they were internalised, creating a sharp contrast between the exterior and interior of his architecture.

19. Kazuo Shinohara: House on a Curved Road, Tokyo, 1978. Interior

20. Sueda Art Gallery, Yufuin, Oita Prefecture, 1981. Exterior

21. Kuragaki House, Tokyo, 1977. Exterior

22. Shokyodo, Toyota, Aichi Prefecture, 1979. Exterior

23. Awazu House, Kawasaki, 1972. Interior

24. Sueda Art Gallery, Yufuin, Oita Prefecture, 1979. Interior

25. Itsuko Hasegawa: Shonandai Cultural Centre, Fujisawa, 1991. Exterior

26. Iida City Museum, Iida, Nagano Prefecture, 1988. Rooftop detail

27. Kuragaki House, Tokyo, 1977. Interior

Externally, these houses employed simple geometric volumes and a meticulous symmetry, especially inside. At first sight, they seem quite unlike the explicitly disjunctive formal assemblages of Hara's first projects. The tone on the outside is resolutely dark, close to black, particularly in the case of the Sueda Art Gallery (fig. 20), with its clapboard-finished walls, reinforced by a dark, pitched and heavily tiled roof. Only the Awazu House (1972) has a flat roof, through which the house is entered. Seeing these buildings in their nondescript urban or suburban environments – especially when one recalls the more aggressive stance of the previous three – one can only be surprised by the quiet modesty or neutrality they project.

Yet this outward simplicity is by no means an endorsement of the surrounding conditions; subtle and not-so-subtle forces of dissimulation are at work and are manifested in the applied details, the combination of materials, limited and small outside openings, expressionless – and often windowless – facades, or no facades at all. The overall impression is perhaps closest to certain rural farmhouses or barns appearing through the filters and distortions of memory. The best examples of such outwardly vernacular architectures in the Reflection House series are Hara's own residence (1974), the Kudo Villa (1976), the Kuragaki (1976) and Niramu houses, but the Shokyodo and Sueda Art Galleries are equally indicative of a more direct native influence on their designs (figs. 21, 22). In 1978, Hara commented on the environment and his response to it:

> In my house or the Awazu House I tried to protect the inhabitants against the physical and psychological interference from outside. If the house has a strong centre, the pattern of interference is automatically reduced ... At the moment we are pessimistic and I envision a future in which the last remnant of the natural environment would be a semblance of daylight.[15] (fig. 23)

Statements such as this can be better understood after one enters these houses. Inside, the visitor encounters something radically different. Everything is white or close to that colour, filled with light, festive or ritualistic, and in contrast to the plain exteriors, extraordinary. The sculpted shapes of the skylights and walls, the stairways, and everything else in the interior, appear to have been brought about by carving out a void within the 'solid' volumes of the exterior forms, as if Hara, like a miner, were working from the outside in; that is to say, *subtracting*. Indeed, a cave-like central space is typical in each of the Reflection Houses and some other small buildings dating from the same time. In Hara's designs, the interiors of these houses are choreographed like religious retreats, sanctuaries, or especially, cave temples. But it is not immediately clear what is being worshipped in them. The answer dawns on the visitor slowly, in time, as one observes and 'listens' carefully. Beyond the feeling of a protective enclosure, two scenarios emerge. Inside the ethereal and solemn space one senses, on the one hand, a stage for nature's fairy tales, and, on the other, memories of urban sceneries. Hara introduces nature indirectly, by way of the sunshine entering through the skylight above the central space that descends the sloping site. Reflected on the light-coloured glassy surfaces of the sculpted interior, the space comes alive, gradually changing from the brightness of a halation to the dimness of dusk, but always luminous and glowing. Thus the formal properties of the space are challenged; they not only get caught up in the shifting spectrum of light, changing the mood accordingly, but also seem, like fog, to dissipate from time to time (fig. 24).

With regard to nature-as-light, Ando comes almost involuntarily to mind; for him, the phenomenon of light-and-shadows is a major architectural 'material'. In his early designs, he too closed the realm of architecture to the outside world, only to slice it open with sharp incisions for the sun to pierce through, projecting the most dramatic and poetic effects. One could also mention Ando's use of courtyards as voids analogous to Hara's central interior spaces. But the similarity ends here; there are important disparities between the two architects. Ando almost exclusively uses reinforced concrete to construct heavy walls as screens on which sharp streaks of light and/or shadows play themselves out in the passing of the day, while the rest of the interior, as in traditional

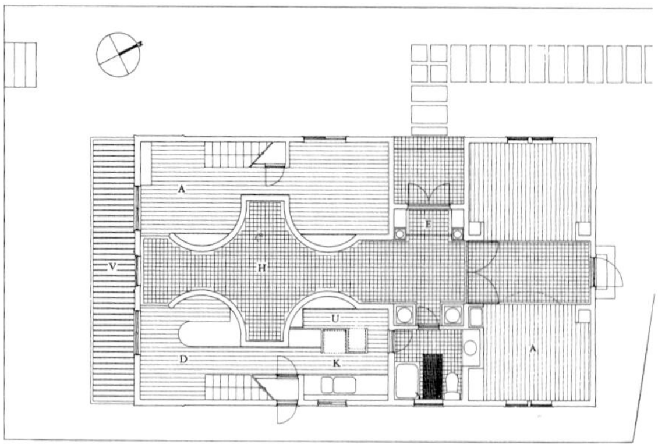

29. Kikuchi Residence (project), 1978. Plan

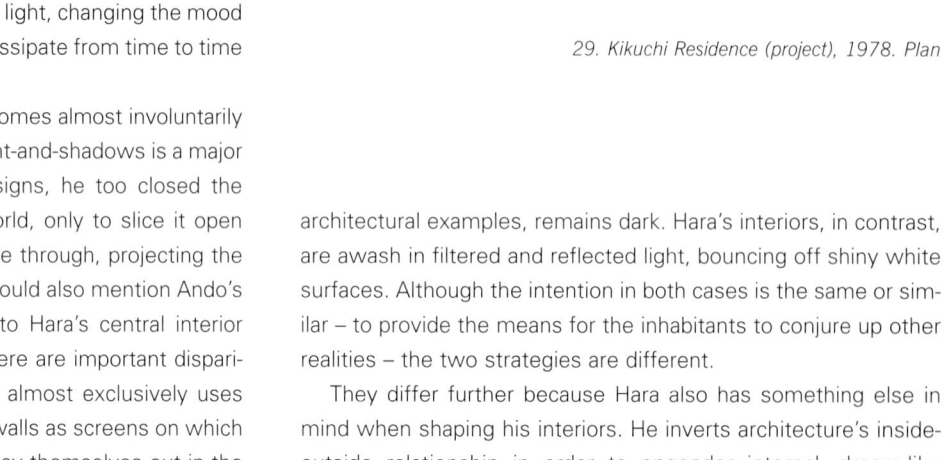

architectural examples, remains dark. Hara's interiors, in contrast, are awash in filtered and reflected light, bouncing off shiny white surfaces. Although the intention in both cases is the same or similar – to provide the means for the inhabitants to conjure up other realities – the two strategies are different.

They differ further because Hara also has something else in mind when shaping his interiors. He inverts architecture's inside-outside relationship in order to engender internal, dream-like urbanscapes. In other words, everything inside is designed so as to suggest that individual spaces or rooms are buildings – with their metaphorical facades, windows, and often translucent roofs – overlooking a common central and cavernous 'outside' space, amply lit from above by skylights. This 'public' domain features stairways, and electric light fixtures that echo urban streets and miniature plazas with street lamps, as best exemplified in Hara's own residence. Here, the realm of architecture becomes the stage of imaginary or fictive urban sceneries resonant with the manifestations of nature. In light of the hopeless outside world, Hara intends, on the one hand, to bring nature as close to architecture as possible, and on the other, to recover and recreate the city by 'embedding [it] within the house'.[16]

In doing so, Hara shares some interpretations of and intentions for nature and architecture with the avant-garde designer, Itsuko Hasegawa, who has declared that she regards 'architecture as another nature'.[17] Perhaps not surprisingly, they have both used certain elements such as undulating cloud-like forms, and habitable outdoor rooftops – a new urban *topos* – created as the extension of the land and the public domain on top of their almost equally scenographic architecture. This is more evident in Hara's later works (figs. 25, 26). The difference between Hara and Hasegawa, however, is that while the latter's architecture is closer to a constructivist (or deconstructivist) paradigm, Hara's is

30. Toyo Ito: White U in Nakano, Tokyo, 1976. Interior

28. Niramu House, Ichinomiya, Chiba Prefecture, 1978. Interior

patently non-constructivist, although the use of certain constructional elements in some of his works, such as the Umeda Sky Building and the Sapporo Dome, are reminiscent of this mode of design.

The initial and most important aspects of Hara's designs are topographical considerations whose origins can be traced back to the memories of his early childhood. One of these relates to the topography of the mountainous landscape around Iida City, located in the Japan Alps, where he grew up during the war. He was fascinated by the phenomenon of deep valleys, particularly the way in which the bordering ranges of mountains reflected light, like mirrors illusionistically reversing the sun's direction. Of course a valley is a natural formation, but Hara readily associates this topography with the configuration of the tightly built narrow streets in Japanese cities. He refers back to this idea in various arrangements within many of his buildings, the largest and perhaps most spectacular example being the atrium in the Kyoto Station complex. Add to this another childhood memory that arguably influences his vision of architecture – his wartime experiences of frequent bombings and the accompanying flashes of light with its effects of halation – and the range of sources becomes more paradoxical and ambiguous. From these contrasting yet merging recollections he has launched his unique system of architectural theories and design strategies. By creating an ambivalent formal type, that is, both valley and street, while revealing its space in unusual light, and further compounding it with other equally elusive meanings, he plays with an increasing determination and sophistication on the ambiguous borderline between reality and fiction.

These additional meanings within the overall ambiguity emerge with more force in some of the Reflection Houses, where the central space is defined by softly undulating surfaces or forms: the Kuragaki House (1977), Niramu House (1978) and especially the unbuilt Kikuchi Residence of 1978 (figs. 27–29). The Niramu House, for example, has certain biological connotations, insofar as the central space – not unlike Ito's White U House in Nakano, Tokyo (1976) – is intestinal or womb-like. But it features light spilling down from above and is less stark than Ito's bent and dimly lit concrete tube (fig. 30). On the other hand, in the Kikuchi Residence, tower-like units populate the high, common space, evoking a surreal village scenery; a vision that can be likened to Giorgio de Chirico's Surrealist townscape paintings. The most haunting images are provided by Hara himself, in numerous magical drawings of what he envisioned when designing the buildings. These fantastic renderings depict imaginary or dream-like scenarios of a long-ingrained cultural landscape, full of scattered traces, as well as 'real' objects such as lanterns, and other marvels. Hara's drawn visions are truly *wonder*ful, enchanted or mythical in conception (figs. 31, 32).

It is testament to the mastery of Hara as a designer-cum-magician that these small buildings within the Reflection House series and others such as the Sueda Art Gallery (1981) or the Stage of Dreams (1982), a cliffside villa on the Izu Peninsula, do actually evoke such wondrous landscapes and sceneries. Moreover – and this may be the real magic of this architecture – the enchantment is sustained long after their completion some twenty-five or thirty years ago. In many ways, they remain beyond verbal description. There should be no doubt that Hara's works of this time are among the most creative not only of his career, but also of New Wave postmodern Japanese architecture.

Nevertheless, the architecture of the New Wave was not without its own internal problems when seen in the broader context from which it emerged. And Hara, like others among the new generation of designers, came to recognise – though also to accept as perhaps the only solution – the contradictions of the reflection and inversion approach in his inward-oriented architecture. He wrote:

31. Kikuchi Residence (project), 1978. Image drawing of interior

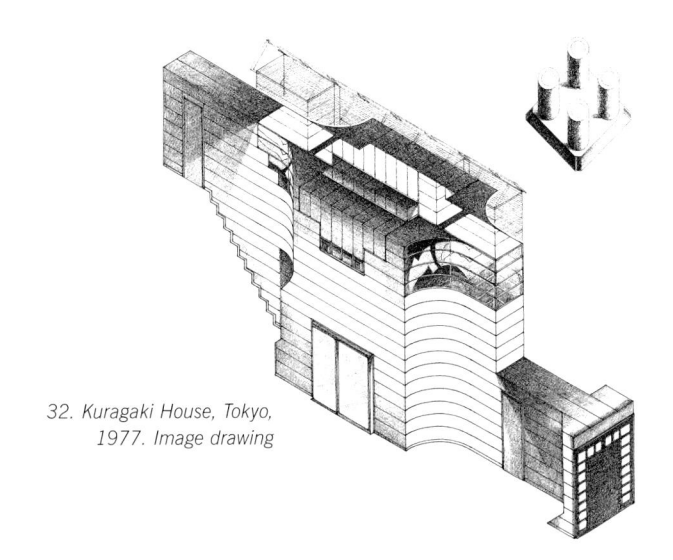

32. Kuragaki House, Tokyo, 1977. Image drawing

What is the social significance of inversion? Simply put, it is the emphatic expression of a self-constituting image of space. Each house has its own centre, and holds every other space within its bounds. Consequently, the term 'social' is no longer useful. Modernism sought the illusion of a pre-established and humanistic harmony, but we have seen how the outcome of 'free planning' was to disrupt and even destroy conventional urban fabric and street scenery. If one wants to realise a harmonious urban fabric according to the tenets of Modernism, some despotic architect must redesign the entire world in some version of Mies's homogeneous spaces. In the dense urban tissue of Japan where there is no space between houses, such a vision for harmony is promptly dispelled. Thus, without the means to create a harmonious and self-sufficient architecture, a paradox arises: an essentially anti-social autocratic method is the most congenial pro-societal alternative.[18]

33. Cityscape of Shibuya with a forest of neon and various media screens at night

34. Nakatsuka House 'Stage of Dreams', Ito, Shizuoka Prefecture, 1982. Interior

35. Sueda Art Gallery, Yufuin, Oita Prefecture, 1981. Elevation

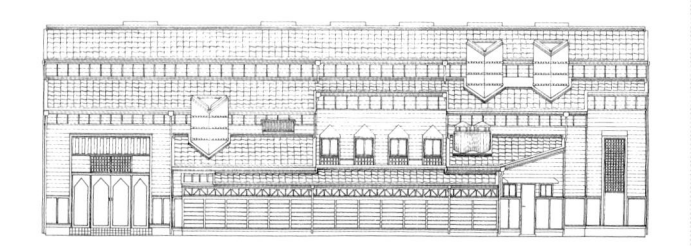

TECHNIQUES OF ILLUSIONS / LARGE-SCALE URBAN PROJECTS

A change in Hara's architecture, mirroring the general trend within the New Wave, was prompted by several external developments in the early 1980s. As the economy began to recover, conditions for architecture and architects gradually started to improve. Apart from some reduction in pollution and congestion, the Japanese city did not change significantly, but there were more opportunities to build. Designers received an increasing number of commissions, many of them large-scale projects. This accelerated to truly unusual proportions with the onset of the 'bubble economy' from the mid-1980s on, when Hara's generation of avant-garde architects not only came of age but were also absorbed, to various degrees, into the establishment. With the transformation of such previous 'urban guerrillas' as Tadao Ando, Arata Isozaki, Shin Takamatsu and others into celebrated superstars, architecture and the city became widely popular and much-discussed topics in the media. The architectural counter-culture assumed the status of the accepted and sought-after standard. As previously noted, this phenomenon was a double-edged sword, producing equal quantities of inferior and superior architectures. But the status of the city was unequivocally raised; its chaotic conditions, ephemerality and state of flux were now regarded as limitless opportunities for creativity. In this change of attitude, one should recognise the primary role played by the shift from an industrial society to an information society; the emerging new economy and the ensuing social developments have increased the range of contradictions inherent in culture, and, by virtue of new information and media technologies, the distinction between the actual and virtual reality of architecture and the urban environment has been further blurred (fig. 33).

Meanwhile, Hara was continuing his research into village architecture around the world, and was becoming involved in large

36. Tasaki Museum of Art, Karuizawa, Nagano Prefecture, 1986. Exterior

public buildings, which would eventually take over from the smaller, private residences. The increased scale and the different nature of these new commissions led to the readjustment of his design strategies while introducing other related aspects of his theoretical investigations. The introverted designs of the Reflection Houses and his opposition to the Japanese city gradually gave way to increasingly open solutions, which were also more sympathetic to the surrounding urban fabric. He continued to search for new ways of exploring the experiential qualities of architecture and the broader environment. The enclosed central space within his residential designs now occasionally took the form of an open courtyard. Other small-scale projects such as the Sueda Art Gallery and the 'Stage of Dreams' or Nakatsuka House were designed as if attempting to realise only one half of his previously symmetrical houses, opening in this way the inner core – the central space in the earlier designs – to the outside court or a view through doorways and glass walls (fig. 34).

These projects emphasise another element in Hara's designs: the buildings are made up of several distinct layers of space with a progression from the front or entrance facades to the higher gallery and living spaces (fig. 35). Beyond the spatial gradation, in-between walls are assigned significant roles requiring an increased attention to the quality of their surfaces. Of course, in his Reflection Houses, the surfaces of the applied materials were never neglected, and a concentrically layered arrangement had already been utilised in the Kudo Villa, but in his works from the mid-1980s on, Hara explores the possibilities of such elements and strategies more extensively, developing them into exquisite forms of art. A new material for Hara in this respect was glass. The first building in which he experimented with the reflective glass surfaces was the Tasaki Museum of Art, followed soon afterwards by a private residence, the Kenju Park 'Forest House' (1987) and the large Iida City Museum, among others.

In both the Tasaki Museum of Art and the 'Forest House', Hara's design features a semi-open courtyard surrounded by the wings of the building. The rectangular perimeter blocks around the courtyard are arranged geometrically according to a straightforward post-and-beam structural system. In the case of the 'Forest House', the vernacular-inspired blocks are solid, while at the museum they are often permeable. Thus in many places they provide a visual, if not physical, filter, offering an ambiguous relationship between inside and out (figs. 36, 37). In both cases, ambiguity is achieved further by means of continuous floor-to-ceiling sashless glass surfaces, which are wrapped around these partially enclosed spaces in order to afford a better view of, and closer relationship with, the courtyard and the landscapes beyond. These extensive glass surfaces are laid out in a sharp zigzagging pattern; a pleated edge, or fractal boundary, that conforms to a 45-degree geometry. This may sound simple and not particularly unusual, but in Hara's hands the result is stunning. Due to multiple reflections and simultaneous transparencies, the division between inside and outside realms is blurred beyond clear recognition; the spatial quality is almost cinematic. The phenomenal experience of moving in and out of these buildings cannot be captured by photographs, which can show only a few of the infinite images and virtual realities one encounters (figs. 38–40).

It is clear that with this new mode of design, Hara aims at bringing together in an inseparable unity both the ambiguity of nature and the ambiguity of perception or human consciousness. In so doing, he finds himself once again in close companionship not only with a number of contemporary Japanese architects, but also with many elements and qualities of traditional and vernacular Japanese architecture. Although his means and solutions are different from those evident in long-standing Japanese traditions, he builds upon the same principles. He experiments with various vague definitions of space – and of just about everything else – whose boundaries are not rigidly established, in this case those between exterior and interior domains. In Hara's efforts to probe architectural meanings and human perception, the most interesting element is his recognition of, and reliance on, two apparently incompatible sources, which he believes lead in the same direc-

37. Kenju Park 'Forest House' Nakaniida, Miyagi Prefecture, 1987. Exterior

38. Tasaki Museum of Art, Karuizawa, Nagano Prefecture, 1986. Reflected images

39–40. Kenju Park 'Forest House' Nakaniida, Miyagi Prefecture, 1987. Reflected images

tion and are therefore equally important in his work: the traditional and the futuristic. The former is grounded in the uniquely phenomenological disposition of Japanese art and culture, while the latter derives from contemporary philosophy, psychology, information theories, science and even the latest electronic and computer technologies. In his subsequent and progressively extensive projects, he blends these sources with no obvious contradictions.

What Hara seeks to achieve in his architecture is a totality of experience based on human perception and consciousness, which he calls 'modality'. In an essay on the topic 'From Function to Modality', he distinguishes the desirable tenets of contemporary architecture as modality as opposed to function, consciousness as opposed to the physical body, and an approach analogous to the workings of electronics as opposed to machinery. He upholds these new considerations in contrast to the preoccupations of modern architecture.[19] Taking such ambitions further, Hara has adopted a growing number of new elements and strategies. The Tasaki Museum introduces to his architecture roof elements shaped as clouds. Wrapped in highly polished aluminium or stainless steel, their glittering and elusive surfaces, while blending into the cloudy skies, undermine a clear reading of their forms. They reappear in the Yamato International, the Iida City Museum, the Green Hall of Musashino Women's College (1990), the Sotetsu Culture Centre (1990) and the Umeda Sky Building, among others.

It is perhaps no coincidence that Hara, like the medieval Zen priest, scholar and poet, Kobo Daishi (774-835), known also as Kukai, uses the analogy of clouds to express ambiguity or uncertainty. As David Edward Shaner explains when discussing the teachings of Kukai, in his texts Kukai 'metaphorically uses such expressions as "mist", "clouds", "smoke", and "dust" to represent the manner in which thetic experience covers our awareness of that which is innate'.[20] A similar portrayal can be observed in traditional Japanese descriptive arts. Paintings intending to avoid a one-point perspective, and thereby the one-way understanding of a conveyed reality, often applied floating golden clouds to different areas of the screen, which variously covered and revealed the depicted sceneries behind, breaking up the continuity and unity of the whole. This technique, called *unkaho* in Japanese, was a means not just to fragment the statement or the intended meaning, but to open it up to broader interpretations and more complex readings. The unseen realm or voids were meant to be filled by the viewer's imagination (fig. 41).

The episodic character of Japanese 'stroll' gardens (*kaiyushiki-niwa*), which were laid out so as to engage the visitor in a game of hide-and-seek (*miegakure*), was the result of similar intentions. While the unrelated elements offered the concreteness and immediacy of engagement, the whole was bound to remain elusive; the observer was therefore forced to play an active role in its constitution (fig. 42). One should also mention here the gardens with 'borrowed scenery' (*shakkei*), in which elements such as trees or a mountain in the distance could be perceived as part of the garden by carefully screening out the zone in between; the experience of the 'whole' was conjured up by illusion.

Closer still to the domain of architecture is the traditional *sukiya*-style residence. Here, definitions of, and/or relationships among, spaces and realms such as outside and inside, centre and periphery, nature and artifice, the finished and unfinished, are fluid, informal and profoundly ambiguous. Instrumental in attaining such flexibility was a wide range of two-dimensional elements: the moveable panels of *fusuma* and *shoji*, the equally mobile *amado* (wooden panels keeping out the rain), *byobu* (portable folding screens) and many other devices. By these means, spaces could be opened and closed, made to interpenetrate, overlap, or wrap around layer by layer in variable configurations. The spatial matrix of this type of residence was not only flexible, but also situational (fig. 43).

Returning to Hara's projects based on the notion of modality, one notes the increased application of 'thin' layers, be they glass

41. Traditional Japanese screen painting showing the technique of unkaho *or 'golden clouds'*

panes, perforated wall planes, or various screens, all with careful attention paid to their surfaces. There is a shift of emphasis in these designs towards multilayered arrangements in which the two-dimensional elements, as much as the spaces in between, tend to overlap and penetrate. Such an approach in Hara's earlier works can be best seen in the Kudo Villa (1976), which features a concentric or nested system of layers. In more recent, larger projects, layering is more linearly sequential and varied. The Tsurukawa Nursery School (1981), for example, is defined by a system of parallel walls punctuated by regular openings, which are revealed by traversing the spaces between and through them (fig. 44). More intricate is the Yamato International Building (1987), designed with a delicate multilayered system of elevation facets like vignettes, which recede from the west to east. The chiselled fragmented layers are covered with meticulously executed, highly polished aluminium panels, which mirror the surrounding environment and reflect the infinite variety of the shifting spectrum of light. The overall impression is vaguely familiar; it appears as if Hara has reversed one of his Reflection Houses, perhaps his own residence. The internalised facades there seem to have been turned inside-out, without relinquishing their previous roles in the process. Hara's notion of 'embedding the city in architecture' returns here with the force of another ambiguity: the carved-away layers of the building emerge as an unreal vision, like the mirage of a hillside settlement straight out of fantasy; Hiroshi Watanabe has called it 'a high-tech Shangri-la'[21] (fig. 45).

Similar layering is applied in both the Ose Middle School (1992) and the Iida City Museum (1988), which are clearly among Hara's most impressive and mature works. The fragmented roof forms, along with the reflective aluminium and other highly polished surfaces, allude to the surrounding mountainous scenery (figs. 46–47). Located on sloping sites, these elongated complexes unfold along sequences of narrow but occasionally permeable volumes and an impressive series of either long courtyards or rooftop terraces in stepped arrangement. In the case of the museum, the freestanding hollowed-out and 'broken' walls on the rooftop promenade and observation platform, overlooking the valley below, suggest an ancient ruin or acropolis (fig. 48). Such a reference is not accidental, since the complex occupies the site of the previous medieval castle compound, whose few remains are preserved and incorporated in Hara's scheme. The metaphor of ruins might also spring from the realm of Hara's subconscious, recollecting his wartime experiences, a common trait among contemporary Japanese architects, most notably Arata Isozaki. These two buildings, seeming to grow out of, and continue the land, present outstanding examples of vertical layering. Other works in which the topography of the land is either continued or recreated are the Tsurukawa Nursery, the Yamato International, the Umeda Sky Building and the Josei Primary School, where the numerous tiled roofs are also involved in the vertically layered spatial system. Appearing first in a rudimentary form at Hara's Keisho Kindergarten, the habitable roof, forming a connected network of flowing, often public, spaces, is another method of interpreting the city as topography.

MODAL SPACE OF CONSCIOUSNESS / TOWARDS AN ARCHITECTURE OF THE INFORMATION AGE

Attempting to elucidate, let alone replicate, the subtle and elusive workings of human perception and consciousness – the ways in

43. Interior of a traditional sukiya-*style residence in Yanai, Yamaguchi Prefecture*

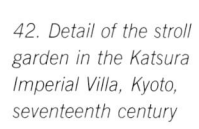

42. Detail of the stroll garden in the Katsura Imperial Villa, Kyoto, seventeenth century

which we read and understand the world – is a formidable task within the reality of material construction. Throughout the years, Hara has explored many strategies to transcend the tangible attributes of architecture and evoke its realm of intangibles. Layering has been one such strategy, along with the choreography of elements for movement through and around them, and the exploitation of the tactile as well as visual qualities of surfaces. Transparency, translucency, overlaying, permeability and reflection have been means actively to involve both body and mind. Hara has shunned the mechanistic model of modern architecture, opting instead for the more flexible informational model, foregrounded increasingly by contemporary electronics and computer technologies. Thus, although constructivism is not one of the dominant features, if at all, of Hara's architecture, he does not hesitate to take advantage of special structural solutions, as seen in his Umeda Sky Building and Sapporo Dome. More importantly, he embraces new technologies. His involvement with electronic technologies began when he started to give more serious consideration to the notion of modality.

Hara's fascinating project entitled *Modal Space of Consciousness* (also known as *Robot Silhouette*), first started to develop when he designed an installation model for an architectural exhibition in Graz, Austria, in 1984. He further refined the design for another exhibition 'Tokyo: Form and Spirit', held at the Walker Art Center, Minneapolis, in 1986. Within this large model, he installed several acrylic panels on whose transparent layers was etched a web of circuit-like patterns. Randomly arranged behind one another in a dark space, the panels were lit according to a computer-controlled program, illuminating the etched patterns with various degrees of intensity, and in different sequences and configurations. As Hara explains, although it was possible to programme and therefore control the sequence of the individual panels, each of which corresponded to a facet of human consciousness, overall control of the superimposed system proved largely impossible and altogether undesirable; the constantly shifting fields retained the ability to flash up an infinite number of unexpected images.

While working on *Modal Space of Consciousness*, Hara began to introduce similar ideas and elements in his building projects. As his architecture became more open, relative to the closed Reflection Houses, the amount of glass surfaces increased, and Hara subjected these to special treatments. We have already seen in the Tasaki Art Museum, the Forest House, Iida City Museum and, to some extent, the interior of Yamato International, the broken or zigzag and curved layout of the glass panes. Another strategy is to etch or imprint their surfaces with patterns. Many windows and glass walls in these buildings feature network-like surface motifs. Often, these take the shape of 'rolling clouds', mirroring those he used on roofs. These clouds appear as translucent areas in the transparent glass, overlaying the inside or outside view (fig. 49). Hara also uses similar motifs as independent elements, like cut-out metal and glass templates, or on details such as handrails, parapets and overhead arches, in most buildings dating from this time.

In the Iida City Museum, these motifs become part of the metal-truss system placed upon a forest of cylindrical concrete posts supporting the roof above the long entrance hall. With this leafy-crowned 'forest' growing inside, the hall is designated as the site or garden of an internalised nature, continuous with the forest and landscape outside (fig. 50). From here it was only one more step for Hara to begin using such patterns on solid walls, ceilings, floors and lamps or lighting fixtures, turning their use into a special form of decorative art. Although occasionally the application of such decorative elements seems overdone, they are not implemented without purpose. Thus, this architecture exemplifies what the Japanese call *tansei* (decorous). Particularly noteworthy are Hara's extensive works in colourful stone inlay; on walls, they recall shadows, or silhouettes or, it might also be said, the friendly ghosts of the otherwise unseen. Sometimes they appear as immaterial demarcations of certain areas, often modulating between reflective and matte surfaces (figs. 51–52). Both inside and out, another kind of floor treatment appears with increased frequency in Hara's designs after the mid-1980s.

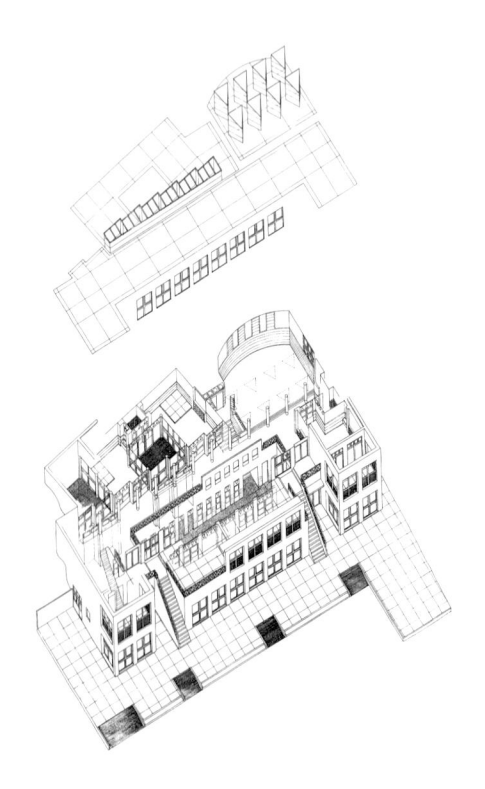

44. Tsurukawa Nursery School, Machida, Tokyo, 1981. Axonometric

Between Reality and Fiction

45. *Yamato International Building, Tokyo, 1987. Exterior view seen from the north-west*
46. *Ose Middle School, Uchiko, Ehime Prefecture, 1992. Courtyard detail*

47–48. *Iida City Museum, Iida, Nagano Prefecture, 1988. Remote view; detail of rooftop scenery*

Recalling traditional stone gardens, where a wide range of surfaces is used to define areas of different tactility, he uses pebbles, tiles, polished and matte stones, water and grass-covered zones for the delight of both body and mind. These works aspire to a true phenomenology of the ground.

In terms of applying electronic and computer technologies, the design and resolution of the huge open atrium space in the Umeda Sky Building comes closest to the *Modal Space of Consciousness* installation. Around and beneath the large corona within the connecting, multistorey sky deck, a host of light sources, including flood-lights, provides moving patterns in an animated show as night falls (fig. 53). In most ordinary buildings, however, there is little opportunity to apply the kind of electronic apparatus required by such a performance. Nevertheless, Hara is a master of the utilisation of natural light. Frequently, the etchings in his glass panels are arranged so as to be highlighted by penetrating beams of sunlight. His Yukian (1988), a small teahouse enclosing two 'traditional' wooden tearooms (*chashitsu*) within a large concrete building, provides one of the best demonstrations of this. Teahouses are designed to conjure up a reality beyond the everyday world, and are places of meditation. The meditative state is induced as much by the architectural qualities of these introspective microcosms as by the rituals of the ceremony itself. It is the range of natural materials, their delicate surfaces and the closed nature of these tiny spaces that define the experience. Insofar as Hara mixes traditional and new techniques and materials to foster the proper frame of mind necessary for the ceremony, the Yukian displays a wider variety of surfaces than usual. Among the elements are narrow vertical incisions in the outside concrete wall, which allow sharp streaks of light to cut into the space.

49. Yamato International Building, Tokyo, 1987. Patterns on etched glass walls

50. Iida City Museum, Iida, Nagano Prefecture, 1988. Interior of the lobby

These bounce around on the shiny surfaces, momentarily hitting etched-glass panes whose illuminated patterns appear as holograms floating in the air, while the interior is also illuminated in a mysterious spectrum of light (figs. 54–55).

Meanwhile, Hara's fascination with various types of valley architecture had not diminished. Many of his recent large complexes have offered him the chance to implement the type in renewed forms, necessitated by the nature or function and size of the projects. In these cases, he needed to consider more extensive structural systems and special technologies. Although small in comparison with such complexes as the Kyoto Station and Sapporo Dome, the Hiroshima Motomachi High School (1999) is one such case. As Japan faces a serious decrease in birthrate and an increasing number of older members of society *(koreika shakai)*, in future years, fewer students are expected to attend school. One of the consequences of this is that there is a growing competition among school districts to attract students. New schools must therefore offer more conveniences and amenities. They must also be multifunctional – often serving as community centres as well, or making provision for this in the future. Using the semi-open space of the multistorey atrium-cum-valley that is continuous with that of the city, Hara once again provided 'urban' elements, among them bridges and escalators, in lieu of, or as well as, regular stairs (fig. 56).

On a scale unprecedentedly large for Hara, in the Kyoto Station Complex he activates the gigantic public space of its atrium in a similar way, through the dynamics of movement and the spontaneous occurrence of events. At both ends of the atrium are series of escalators, along with a wide cascading stairway at the west of the space, all leading eventually to various rooftop plazas accessible by the public. Replete with eateries, cafés, information centres and many other services and activities, and outfitted by a broad range of design elements or 'attractors' – including the well-camouflaged mechanical equipment and machinery, permanent displays of art and media – this magnificent futuristic atrium is an exhilarating, almost mythical space (fig. 57). Contiguous at many points with the outside world, it is a rare achievement in contemporary urban architecture.

Such qualities, one has to acknowledge, can be attributed in part to the impressive dimensions of the station. Paradoxically, while this holds true of the building's interior – a small city itself –

Between Reality and Fiction

of the participants in the international competition for the project, to reduce the volume of the building. In fact, one of the reasons why Hara's scheme won the competition was that it offered a better overall solution to this problem than those of his competitors. As can be seen from the results, while the city gained a splendid public space and many new facilities, it also further compromised its urban charm.[22]

The same fate has not befallen Hara's other exceptional and equally gigantic project, the Sapporo Dome, for several reasons. It is located on the outskirts of the city, in a large, park-like area, closer to the natural landscape. Moreover, due to the nature of this multi-use sports and entertainment complex, Hara had a freer hand in shaping the Dome. He chose a simple bubble-like fluid form wrapped in stainless-steel sheets. Shape and material join forces here to create the illusion of a smaller structure; its enormous volume seems to dissipate like mist as it reflects the changing light conditions (fig. 58). The other marvel of the project, beyond space and architecture, is the futuristic technology that allows the 90 x 120-metre playing field of natural turf to be moved in and out of the Dome. Not only does the superstructure float in a virtual sense, but also the field does so in reality, like a hovercraft on a thin air cushion.

Throughout his many projects, the scale of Hara's architecture and of his thinking and consciousness have been expanding, encompassing and shaping both internal and external horizons, as well as sizes and sites as varied as small internal spaces and vast external worlds. His course of development as a scholar and an

51. Umeda Sky Building, Osaka, 1993. Interior detail of lobby with reflective surfaces

the same cannot be said of the architecture of the complex as a whole. Its enormous external volume and excessive height, relative to the much lower profile of the historic urban environment in which it is situated, have attracted controversy and criticism. Matched only by the similarly enormous University of Tokyo Komaba Campus Buildings (2001), more than any other in Hara's productive career this project proves that architecture, for better or for worse, is necessarily caught up in and shaped by the vicissitudes of economic and political forces. Despite strong popular protest in Kyoto, the municipal authorities, bowing to the pressures of the private corporations – a large department store, a hotel and many other businesses, all wanting to be part of the profitable project – lifted the previously enforced building-height limit of 31 metres, doubling it to 60 metres. With such a formidable programme, there was little chance for Hara, or for any

52. Ueda Shokai Guest House, Tokyo, 1992. Stone pattern on lobby walls

53. Umeda Sky Building, Osaka. 1993. Evening view of atrium

architect has paralleled that of our age, in which progress in the arts, sciences and technologies, productive capacities and knowledge, if not always wisdom, have opened new dimensions in both the microcosmic and macrocosmic realms. Combining cosmic visions with cutting-edge technology, Hara has also ventured into outer space, if not in reality, then by force of his imagination. Starting with 'gazing' at the sky through the aperture of the corona in the Umeda Sky Building, he has continued to devise colonies for space dwellers and those intending to inhabit the Moon and other planets in the future. His project Extraterrestrial Architecture (1992) probes the limits of both architecture and human imagination, touching a realm that for the moment has to straddle the real and the fantastical, even the metaphysical. On the other hand, in the case of the Miyagi Prefectural Library, Hara brings this idea down to earth; the building is shaped as an 'unreal' vision, like the mirage of a UFO that has just landed in the lush outskirts of Sendai (fig. 59).

Paradoxically, Hara's latest projects – the tiny Cube Houses – may represent a scale as vast as, if not the cosmos, then almost the entire country. Scattered across different regions in Japan, these four, tower-like buildings have no opportunity for any physical connection or relationship with each other. However, like landmarks, or rather, attractors in a huge semiotic field, they are integral parts of a nexus inscribed on the land by the all-encompassing invisible network of communications, or 'spaces of flows', that is created and cultivated by the dynamics of Japan's information society. Hara's designs represent a novel idea, somewhat similar to Isozaki's project for the Kumamoto ArtPolis. Conceived in the mid 1980s, Isozaki's idea was to build a community, a new *polis*, whose members are physically dispersed throughout the Prefecture of Kumamoto, but are linked by being part of the same programme, by the map on which their location is indicated, by unpredictable trajectories, and by the memories of those travellers who have visited the individual buildings and locations. Hara's scheme, however, has another dimension: the Cube Houses are designed according to a mathematical scale of size gradation, and are virtually connected by the workings of our knowledge and memory of such architectural disposition (fig. 60).

Turning this concept upside-down or, more precisely, inside-out, Hara is now at work on another series of small residences. Here, it is the courtyards – whose empty spaces are negative or 'void attractors' – that will form a geometrical scale of gradation. In these projects, Hara continues to pursue on the one hand his

54-55. Yukian Tea House, Ikaho, Gumma Prefecture, 1988. Interior views

56. Hiroshima Motomachi High School, Hiroshima, 1999. Detail of atrium

long-cherished dream of creating metaphorical 'celestial gardens' – implied in so many of his previous architectures – and on the other his experimentations with human perception, imagination and new architectural realities. They bring him ever closer to his desired aim, in which the autonomous architectural object is actively challenged, space is always only imminent, and a sense of place is significantly redefined. Here, place is (re)enacted, that is to say, it is constituted more by the transitory events taking place there, than it is determined by the physical attributes or material disposition of permanent constituents. Increasingly, Hara is operating in that modality, or new mode of architecture, whose possibilities Ignasi de Solá-Morales has so outlined poignantl:

> The places of present-day architecture cannot repeat the permanencies produced by the force of the Vitruvian *firmitas*. The effects of duration, stability, and defiance of time's passing are now irrelevant. The idea of place as the cultivation and maintenance of the essential and the profound, of a *genius loci*, is no longer credible in an age of agnosticism; it becomes reactionary. Yet the loss of these illusions need not necessarily result in a nihilistic architecture of negation. From a thousand different sites the production of place continues to be possible. Not as a revelation of something existing in permanence, but as the production of an event.[23]

QUESTION AND/OR ANSWER / IN LIEU OF CONCLUSION

Lastly, some important questions need to be raised regarding Hara's entire architectural enterprise within a 'floating world'. Is his world of ambiguity a fight against, or a flight from, our present day and its increasingly simulated realities – which are not without some invisible yet powerful controlling forces? Is this ambiguity an affirmation or negation of the conditions in which he operates, or can the meanings, or rather, *should* the meanings of his architecture always be 'undecidable'? Does this 'floating world' of his make us forget or prompt us to remember? If Hara's architecture is likened to the realm of the traditional teahouse, or by extension to the *sukiya*-style traditional Japanese residential architecture,[24] as it can and should be, then these questions can be answered with yet another ambiguity: with the word 'both'. The teahouse has always been a world into which participants in the ceremony try to escape from the troubled and harsh realities of their everyday lives. But at the same time, the 'primitive' yet delicate simplicity of this meditative world, in contrast to the power, hedonism and overly decorative, opulent and even garish architectures in the lives of the rulers at large, has also been the manifestation of a resistive or critical stance. We also know that today, in our all-encompassing, global world of corporate and commodity cultures and apparently apolitical and ideology-free reality, only the slightest discrepancy can exist between an affirmative and resistive position.

57. Kyoto Station Complex, Kyoto, 1997. Interior of atrium

With regard to some of these issues, particularly about the decidability or undecidability of meanings, and the role of language within them, Terry Eagleton has this to say:

> Meaning ... becomes 'decidable', and words like 'truth', 'reality', 'knowledge' and 'certainty' have something of their force restored to them when we think of language rather as something we do, as indissociably interwoven with our practical forms of life. It is not of course that language then becomes fixed and luminous: on the contrary, it becomes even more fraught and conflictual than the most 'deconstructed' literary text.[25]

But it is perhaps Vladimir Krstic, referring directly to Hara's work, who has expressed it best:

Hiroshi Hara

The tenuous line that separates existence from non-existence, along which Hara's ... work unfolds, seems to denote a subversive twist of the simulative hyperreality of the city which, within the hopeless stillness of its world, promises to open up to an entirely new realm of phenomenal and existential dimensions of architecture and in which the mythical poeisis of building appears as a not yet extinct possibility. Yet, [his] work can still be criticised as much as the activity of the seventeenth-century sukiya masters for example, who were accused of overindulgence in the aesthetic matters of architecture as a form of political escapism. However, if the alternative to [his] architecture, relative to the problematic of the Japanese city, is a looming corporate totalitarianism ... then not only does the choice become self-evident, but at the same time the work of Hara ... attains an important political perspective.[26]

If so, although there are no clear or easy answers to the questions above, Hara might have found a fitting response in his architecture to the fluid reality of the world-in-flux in our age of information.

60. Two of the four Cube Houses, Shimabara, Nagasaki Prefecture, 1998

59. Miyagi Prefectural Library, Sendai, 1997. Detail of stainless-steel roof

58. Sapporo Dome, Sapporo, Hokkaido, 2001. Exterior view

NOTES

1. Henry Plummer, *A+U, Architecture and Urbanism Extra Edition – Light in Japanese Architecture,* June 1995, p 242. The 'Floating World' (*ukiyo*) in the title is also a reference to the popular urban culture of transitoriness and make-believe that flourished in Japan during the Edo Era (1600–1868). It grew out of a situation in which townspeople and the merchant classes accumulated wealth, but were ranked low in the social hierarchy, having no political power, which was concentrated in the hands of the shogun and the military (*samurai*) class. They found solace in worldly pleasures and the realm of the imagination, in arts such as theatre, pictorial arts (*ukiyo-e*), poetry and fiction.
2. Motto of the National Public Radio (NPR) programme, 'New Dimensions'.
3. I am referring to the emergence of 'structural' anthropology, and especially to the pioneering work of the French Claude Lévi-Strauss, who conducted his research on the kinship, social life, and system of myths of indigenous peoples; he published his findings in such books as *Tristes Tropiques* (1955), *Structural Anthropology* (1958), the *Scope of Anthropology* (1960), *The Savage Mind* (1962), and several others.
4. All three of these architects received their M.Arch degrees from Harvard University; Maki in 1954, and Takeyama and Taniguchi in 1964. Due to the war, many leading European modernist architects, such as Mies van der Rohe, Walter Gropius, Josep Lluís Sert, Marcel Breuer, Erich Mendelsohn and others, went to the United States and continued their practices there. Many of them also taught in universities – Gropius and Sert at Harvard GSD.
5. Maki graduated in 1952, while Hara received his Bachelor of Architecture in 1959, M.Arch in 1961, and Ph.D degrees in 1964, all from the University of Tokyo.
6. Hara's trips were as follows: 1972 – the Mediterranean; 1973 – Brazil, Peru, small Japanese islands; 1974 – South and Central America; 1975 – Eastern Europe and the Middle East; 1977 – Iraq, India, Nepal and Indonesia; 1978/79 – East Africa; 1980 – Japan; 1981 – Kuwait, Saudi Arabia and Egypt; 1984/88/89 – China; 1989 – Venezuela; 1990 – Indonesia; 1991 – Papua New Guinea; 1992 – Mexico and Japan; 1993 – Indonesia; 1994 – South America, China and Italy; 1995 – Northern Europe and Morocco; 1996 – China; 1997 – Yemen; 1999/2000 – Uruguay.
7. Hiroshi Hara, *Learning from Villages*, Showa Women's University (Tokyo), 1999, p 6.
8. Hara, 'Yukotai Theory 1968', in *GA Architect 13 – Hiroshi Hara*, ADA Edita (Tokyo), 1993, p 32.
9. Hara, quoted in Chris Fawcett, *The New Japanese House*, Harper Row (New York), 1980, p 85.
10. David Stewart, 'The Intelligence of the Senses: A Primer of Hiroshi Hara's Phenomenological Space', in *GA Architect 13 – Hiroshi Hara,* op cit, p 11.
11. Stewart, quoting from Christopher Alexander, ibid, p 11.
12. Hara, *Learning from Villages*, op cit; and in 'Learning from Villages: 100 Lessons – 1987', in *GA Architect 13 – Hiroshi Hara*, op cit, p 88.
13. Robin Boyd, *New Directions in Japanese Architecture*, George Braziller (New York), 1968, p 32.
14. Fawcett, *The New Japanese House*, op cit, p 94.
15. Hara, 'An Interview with David Stewart', in *Architectural Association Quarterly (AAQ)*, vol 10, no 4, 1978, p 27.
16. Hara, 'Reflection and Inversion 1978', in *GA Architect 13 – Hiroshi Hara*, op cit, p 61.
17. Itsuko Hasegawa, 'Architecture as Another Nature', in *AD, Architectural Design Profile No. 90, Aspects of Modern Architecture*, 1991, p 14.
18. Hara, in *GA Architect 13 – Hiroshi Hara*, op cit.
19. David Edward Shaner, *The Bodymind Experience in Japanese Buddhism*, State University of New York Press (Albany, NY), 1985, p 123.
20. Hara, 'From Function to Modality 1986-1993', in *GA Architect 13 – Hiroshi Hara*, op cit, p 128.
21. Hiroshi Watanabe, *Amazing Architecture from Japan*, Weatherhill (New York & Tokyo), 1991, p 38.
22. Unfortunately, much of Kyoto's traditional urban fabric has already been destroyed, gradually undermining the appeal of this ancient capital. Due to slow and inconsistent preservation policies, scores of attractive townhouses *(machiya)* have been demolished, and many of their sites turned into parking lots. Moreover, there is now a growing number of highrise buildings even in the still low-profile and more traditional Kawaramachi-Oike district, such as the twenty-storey high Kyoto Hotel.
23. Ignasi de Solá-Morales, *Differences*, quoted in *Harvard Design Magazine*, Fall 1997, p 2.
24. *Sukiya*-style residential architecture evolved under the influence of the informal teahouse (*chashitsu*) architecture in the late sixteenth century.
25. Terry Eagleton, *Literary Theory: An Introduction*, The University of Minnesota Press (Minneapolis, MN), 1983, p 147.
26. Vladimir Krstic, 'Stillness of Hyperreality: The In(de)finite City', in Botond Bognar (ed.), *Japanese Architecture II. AD, Architectural Design Profile*, no 99, 1992, p 27.

1. Reflection / Embedding
by Hiroshi Hara

I was brought up in a deep valley, surrounded by 3,000-metre-high mountains. The tallest are covered with snow in the winter. In a valley, due to reflection, the sun's direction appears to be reversed. In the morning, I saw the western mountains shining with sunlight and in the evening the eastern mountains changing their colours minute by minute. The mountains, which act as screens reflecting sunlight, and the topography of the valley have deeply affected my architectural design.

Living in a valley, it hit me that light becomes visible only through reflection. The reversal of the view reminded me that a valley is a fictional device created by nature. The mountains reflect light like two rows of facing mirrors. With the valley between, there is a distinct symmetry.

Soon after starting my career as an architect, I began to envision my 'Valley Architecture'. The Reflection Houses such as Awazu House (1972), Hara House (1974) and Niramu House (1978), as well as Kyoto Station (1997) and the Sapporo Dome (2001), now under construction, all belong to this Valley Architecture. I also see the Umeda Sky Building (1993), two skyscrapers connected at the top, as another example of this kind of architecture: my interest is mainly in the space between the two towers, sandwiched by a pair of tall curtain walls. The Miyagi Prefectural Library is what I would call 'Bridge Architecture' in terms of the exterior view, but this is also inspired by valley topography, since bridges span valley-type formations.

Topography is the most important morphological factor in architecture. As a university professor I have undertaken much field research in villages throughout the world. The forms of the villages depend on, and are differentiated by, their surrounding climates and the obtainable materials. However, we must be aware that topography, which is a microscopic geographical condition, also shapes architecture. In this sense, architecture is

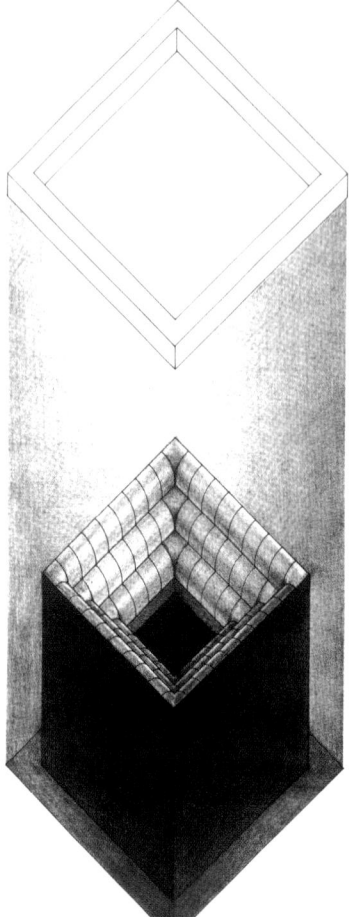

Embedding

topography-designed. When built on slopes, structures resemble each other in the sense that they make a new topography under the specific constraints and possibilities of the slope.

If architecture is designed topography, the city, which is a collection of buildings, is another topography that can be considered an integration of micro-topography. The topographical aspect of the city most typically appears in the streets – valleys between the two facing street facades.

When two surfaces face each other, an axis, a formal structure, is generated. This may also be called symmetry, or more generally, 'facing structures'. Therefore we can speak about valley topography even without mountain ranges: it can be brought about by any two facing surfaces. Two people standing on either side of the valley will meet each other due to the character of the facing structure. Thus by employing this structure, Valley Architecture induces people to meet. We can see this kind of topographical device in ancient amphitheatres and many ritualistic buildings. Clearly this is a factor common to fictionality. Even in today's streets we can take advantage of the valley's fictionality if the facing structure is well considered.

The reason why the city environment offers excitement and vitality is that it often featured the facing structure, or reflection structure. Whenever people get to know each other, this essentially takes place within a facing structure. And to face somebody is to see oneself through the other person's existence. Urbanism, with its streets that work as valleys, literally induces many encounters.

In the Reflection Houses, my concept was to place a valley, designed like a street, in a small house. This process of enveloping valleys or streets in houses can be referred to as 'embedding'. Basically, embedding is a geometrical operation since one geometric form is placed within another. When two forms are alike, they are characterised as 'self-similar'. In the Reflection Houses, however, self-similarity was not pursued. The focus was to embed the valley/street in each form. On the other hand, since all these houses stand on slopes so that in each house a smaller valley is embedded in a larger one, it could be said that they have self-similarity.

Placing a valley or street in a small house is a purely fictional and metaphorical manipulation, since large is inserted into small. The original concept for the Reflection Houses was based on the beautiful *Arabian Nights* story 'Tale of a Fisherman'. It tells of a fisherman who finds a pot in his net. When he pulls out the stopper a giant demon emerges. The fisherman must use all his wit to put the demon back into the pot. This story can be extrapolated metaphorically to the concept of embedding a whole city in a small house.

When I began working on the Reflection Houses more than twenty-five years ago, I never imagined that I would go on to design Yamato International (1987) and the Kyoto Station (1997). But these buildings are also based on the concept of embedding – 'a village within a building' and 'a city within a building'. Though still metaphorical, the embedded elements here come closer to the actual size of villages and cities.

The concept described above is most clearly expressed as form in my own house. However, the other houses, except for the Kudo Villa, basically share the same format, which can be described as follows:
a. The house's public space descends via stairs that take into consideration the site's slope, with the living room at the bottom.
b. The rooms, each with a toplight, are symmetrically located along the public space.
c. Skylights mark the axis created by a and b.

In my own house, each room has an acrylic ceiling referred to as the 'second roof'. This emphasises the distinctiveness of each as a small, independent building. Thus the interior space resembles a street.

Another aspect that interested me was the correlation between form and the interior atmosphere – that is, sunlight and interior brightness, temperature, air movement, resonance, the optimal viewing angle etc, at each point in the house. The situation in which these phenomena take place is called the 'field' in physics. In these terms, the house is an overlay of fields of brightness, temperature, air movement, resonance and optimal viewing angle. Though difficult to predict, the residents' field of movement is of course the first to be considered. Houses that share the same form are convenient when comparing the attributes of these fields, at the same time demonstrating the efficiency and limitations of the form. In architecture, the floor, ie, topography, is an important field, considered as the 'base'. Other fields are phenomena that take place on this base. In the Reflection Houses, the fields have valleys as their base.

The system of the Reflection Houses makes designing the fields easy. With the field of brightness for example, the axis area with the skylight above becomes the brightest, with some transitional zones around it. Designing architecture as fields of brightness has been my basic approach.

'Symmetry' is another term for reflection. If elements are placed symmetrically, the building becomes ritualistic. Here, geometry is concentric, with the axis as an elongated centre. At the time of the Reflection Houses, I was researching the concept of the centre of a geometrical form and I named the axis, ie, the centre of a closed curve form, the 'ridge'. The ridge is obtainable in any closed curve in terms of symmetry. If you strew sand over a closed curve, a mountain-like topography will materialise, and a ridge will appear as a set of singular topographical points. This set of points can be referred to as the 'repeller'. If this topography is turned upside-down, the ridge becomes a valley, which is a negative ridge. The valley, in contrast to the ridge, is a set of points that can be termed the 'attractor'. My fondness for valleys grows from my fondness for the attractor.

Later, the concept of the attractor proved useful within the idea of the 'semiotic field'. When I was designing the Reflection Houses, I had not yet discovered the semiotic field, and was interested only in describing the topological aspects of the valley. Thinking back, I can now say that my concerns with the symmetry and geometry of the valley were a way of turning a house into a spatial attractor.

In the days of the Reflection Houses, the emergence of skyscrapers in Tokyo was establishing a city centre. In contrast to such big centres, I wanted to make houses small centres. For this reason, I brought ritualistic symmetry into their design. In terms of urban form, this idea is close to that of the old cities, or medinas, in the Maghreb countries, where courtyards serve as the centre of the houses, thus making each house independent. The internal valleys and symmetries of the Reflection Houses match the courtyards of medina houses.

However, my thinking concerning symmetry was different. For example, in both the Awazu House and my own residence, the skylight runs from north to south. Strictly speaking, the sunlight entering from the skylight projects a symmetry of light and shadow only around noon. The rest of the time, the sunlight actually breaks the symmetry. In this sense, the house works like a sundial. These constant changes in the internal conditions and symmetry of the house are very important in urban buildings in order to convey the presence of nature, materialised in light and shadow. The symmetry/valley approach makes this visible. The process whereby architecture changes its state from time to time – modality – would become my major architectural theme.

What I found significant was not simply the ritualistic and concentric aspects of the buildings, but their changing state in accordance with the cycle of nature. This is the same as thinking of architecture as a field, ie, not considering it as 'things' but as 'events'. With the field, things are not important in contrast to

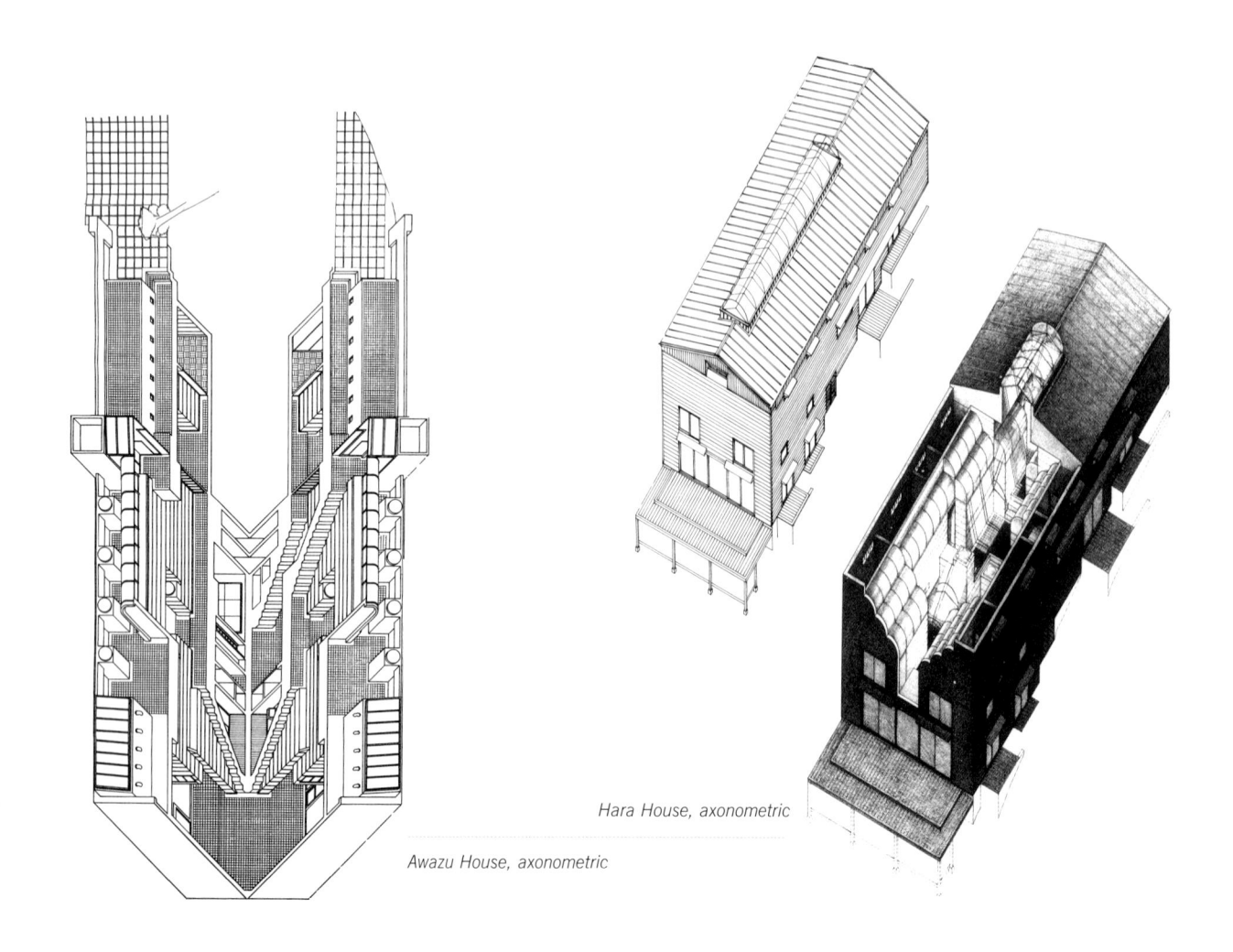

Awazu House, axonometric

Hara House, axonometric

events. In the field of light and shadow, things disappear. With this process of diminishing things, architecture as event – in other words, architecture as space – comes into being. I consider this shift from thing to event the main theme of architecture in the latter half of the twentieth century, on which Ludwig Wittgenstein speculated in his book *Philosophical Investigations* (1953).

Among the Reflection Houses, Kudo Villa (1976) has a slightly different character, though also built on a slope. A small weekend residence in the mountains, it is a place where one can immerse oneself in nature, and is not intended for permanent living. It has no valley embedding but is still based on symmetry. A small flat base is made using *tatami*, the traditional Japanese floor material. Sunlight enters through louvres located on the walls, as opposed to a skylight. In fact, this house is based on 'absorption', which is the antonym of 'reflection'.

Reflection and absorption are opposites in terms of light. In reflection, light comes in through the skylight and is then diffused. In such cases, the interior colour is white and the material of the second roof is translucent acrylic plate. In absorption, light first enters through the louvres and is filtered through glass and paper doors (*shoji*), dimming the interior space. The colour is almost black except for the straw of the *tatami*.

The contrast is also apparent in the field of acoustics. In the reflection type, resonance is deep, while in the absorption type it is shallow. Sound, including the human voice, is literally reflected or absorbed in each type.

Thus reflection and absorption are contrastable concepts of modality. These two modes, though no one has clearly stated the fact, are part of a longstanding Japanese tradition. The mode of reflection can be applied to temple halls containing Buddhist statues, whereas the mode of absorption can be seen in traditional farmhouses. In eastern Buddhist philosophy, the mode of reflection parallels that of *shogon*, a translation of the Sanskrit words *vyuha* (well arranged) or *alamkara* (well ornamented). On the other hand, the mode of absorption corresponds to *jakumetsu* or *jakujo*, which in Sanskrit are *santa* or *sama*, meaning stillness or quietness. Japanese medieval aesthetes attempted to express both of these at once, aiming for the absolute unification of contrary concepts. This results in multiple meanings and ambiguity – typical of the Japanese tradition.

In designing the Reflection Houses I wished to extract each of these two modes, which exist as an amalgam in the traditional Japanese context.

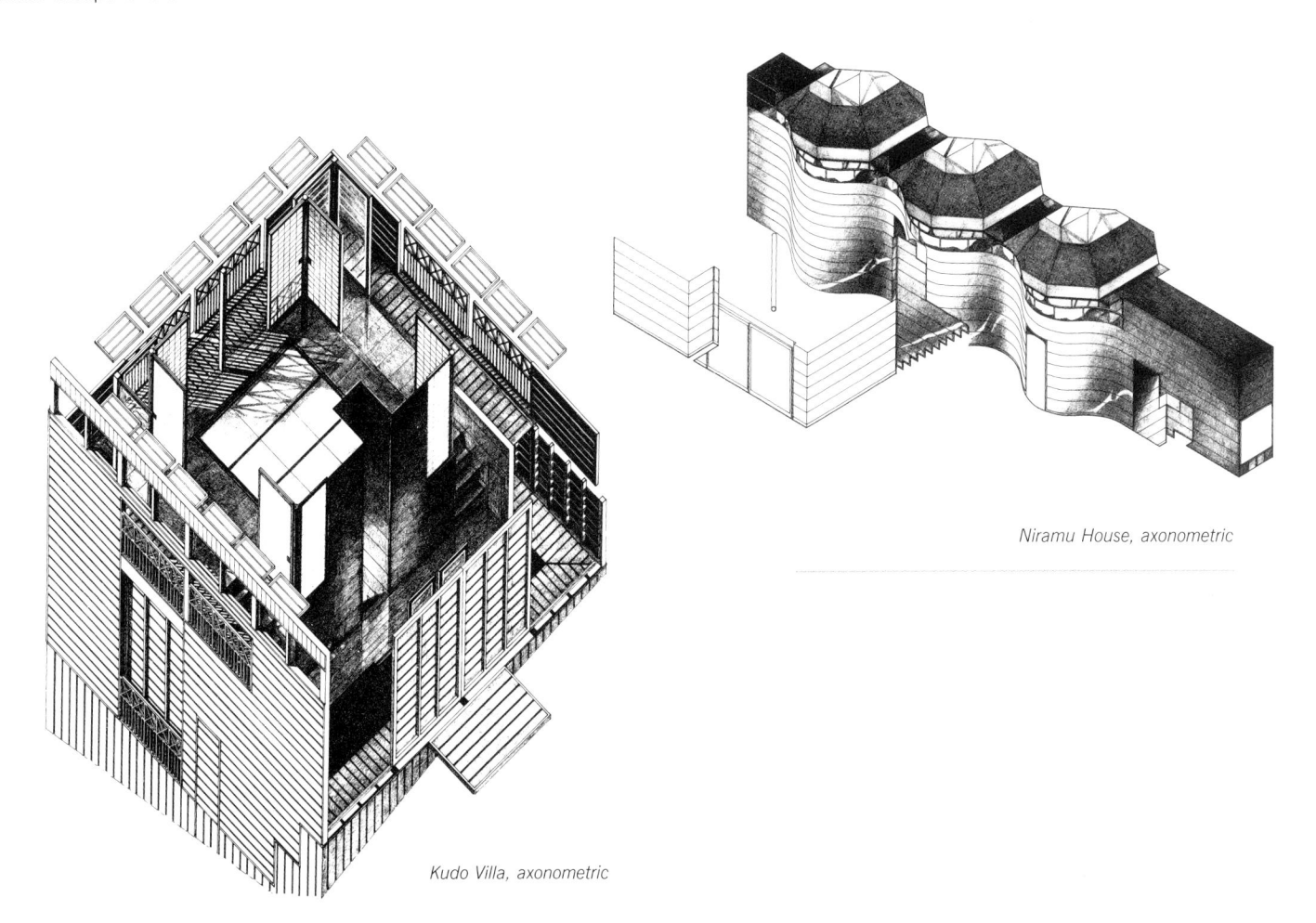

Niramu House, axonometric

Kudo Villa, axonometric

Awazu House, Kawasaki, Kamagawa Prefecture, 1972

This was the first of the Reflection Houses, a series of residences designed on a symmetrical plan. The site is the upper slope of a hill in a Tokyo suburb. Other houses in the series are also built on sloping land. The cheaper cost of land on hillsides probably accounts for the ubiquity of such sites. Slopes gave me an understanding of the significance of topography, and led me to develop the specific style of this series.

It was in the Awazu House that I first employed the technique of framing a part of the site with a rectangular enclosure and arranging the individual rooms within that enclosure to conform to the slope of the site. I adopted a symmetrical plan for two main reasons. Firstly, because symmetry, with its association with Classicism, had been avoided by modern architecture; the adoption of symmetry was an expression of my ideological rebellion against Modernism from around 1960 to 1970. Secondly, I wished to turn the interior of the house into a valley-like space. Symmetry was also a way of endowing the house, small as it was, with a powerful centrality. This was later to become the basis of my concept of 'discrete semiotic fields'; that is, the idea that a discrete arrangement of small centres (composed of groups of houses) is a better structure for a city than a single large central district surrounded by a marginal area. I remain confident of the validity of this idea of discreteness.

At the time, I was intent on developing a new residential form, and I used the same style in subsequent houses. I felt that as long as the form was the same, it did not matter what materials or structure I used, whether wood, steel frame or reinforced concrete. I therefore came to employ many different materials and structural systems not only in the Reflection Houses but in my work in general. *HH*

Homage to the Pillars of Hercules

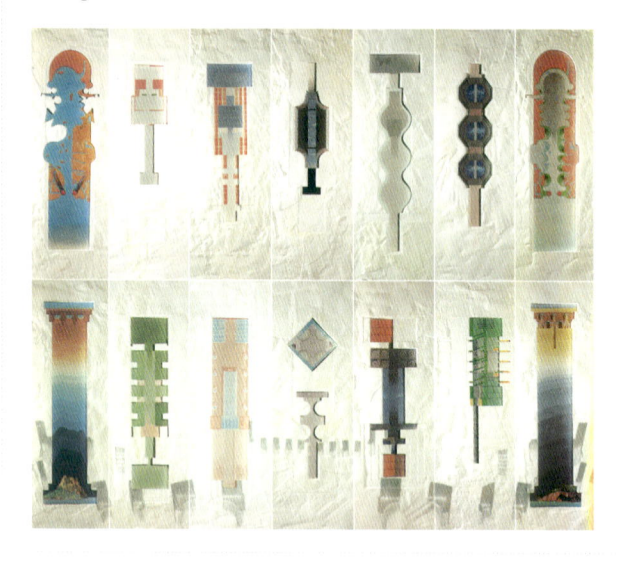

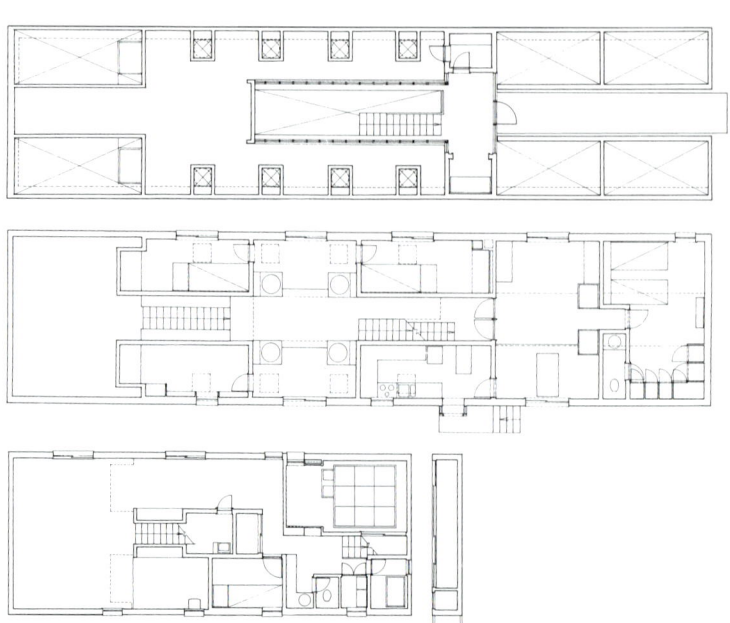

Roof, second-floor and first-floor plans

Opposite: The second-floor hall and first-floor atelier seen from the entrance

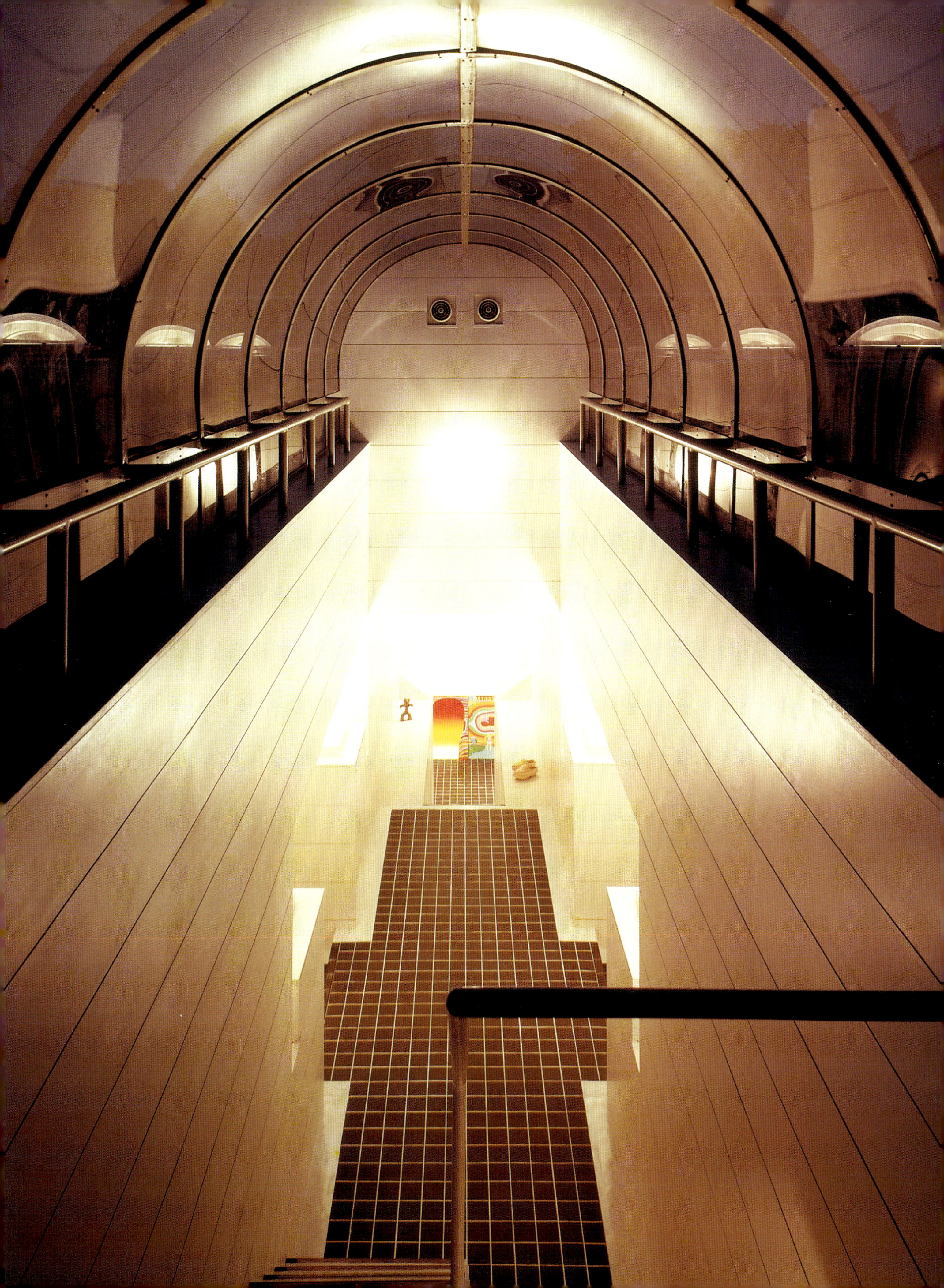

Hiroshi Hara

General view from the west (entry approach)

Awazu House

The high-ceilinged atelier; the study is on the right of the staircase and one of the children's rooms on the left

Hiroshi Hara

Hara House, Machida, Tokyo, 1974

This is my own residence, and has the same form as the Awazu House, but the structure is wood. Nearly thirty years have passed since its construction, but little has changed in the house. It is the most representative work in the Reflection House series for the following reasons. Firstly, the area framed by the rectangular outer enclosure has been clearly endowed with a valley-like topography. Secondly, the house is a clear expression of what I call a 'nest structure' (smaller 'houses' are nested within the house like Russian dolls) as well as my idea of 'embedding' a city inside a house. Thirdly, the exterior of the house is simple in appearance, though this is difficult to see because of the trees, and the true facade is found inside. That is, the design is successful in expressing a reversal of inside and outside.

One simple detail was crucial to the design of this house: a 'second roof' of acrylic panels installed in each of the rooms. At the time, I was interested in acrylic, which was still a new material. A simple method for installing the panels (with clips and screws) was developed. This second roof serves as a lightweight, cloud-like ceiling. As a result, each room in the building looks like a separate house. It is as if a street of houses has been embedded inside the building. Light enters each room via a window, but a skylight above the central passageway also illuminates the room through the second roof. The all-white interior is brighter than the outside garden, which is shaded by trees. This brightness underscores the idea of reversal as well as the concept of a 'small centre'. When the lights are on in the rooms at night, the second roof of acrylic panels glows, and the rooms themselves become sources of light.

The second roof gives autonomy to each room. I wanted to express in this way a certain relationship between the family and the house. I later discovered a similar relationship in a compound in Africa, where each family member was treated as a distinct individual. HH

Upward view towards the entrance seen from the living room

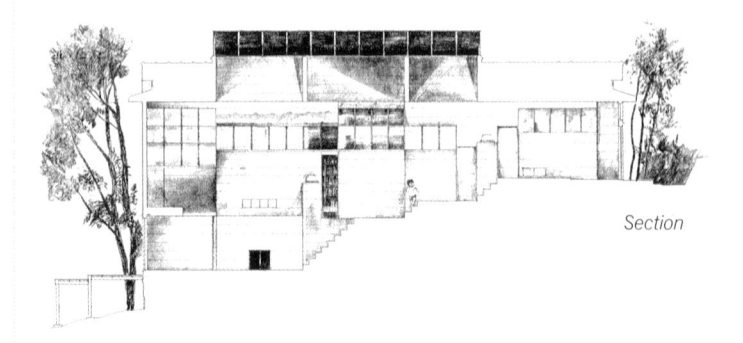

Section

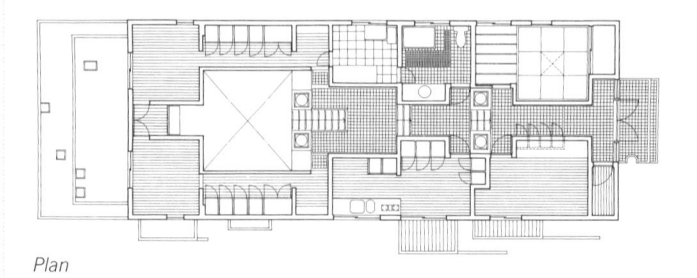

Plan

Opposite: View towards the entrance from the high-ceilinged living room at noon

Overleaf: View of 'second roofs'

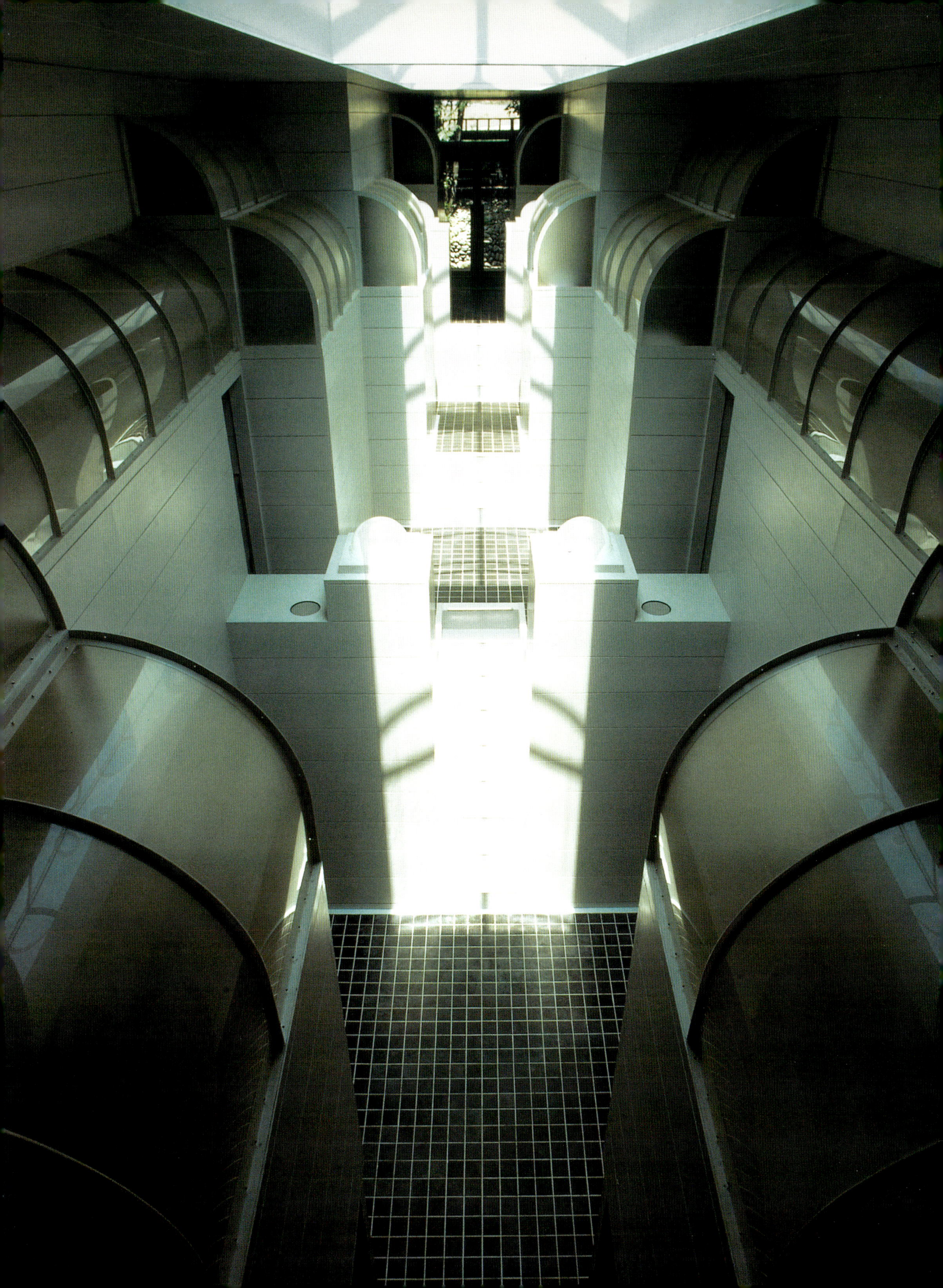

Hiroshi Hara

Hara House

Kudo Villa, Karuizawa, Nagano Prefecture, 1976

This small villa was conceived both as part of the Reflection House series and as the polar opposite of others in the group. If 'absorption' is the opposite in modality of 'reflection', then this is an example of an 'Absorption House'. It is similar in some ways to a traditional Japanese house, except that the plan of this villa is symmetrical.

Through the use of louvres on the outer wall, as well as sliding doors and suspended panels with louvres, I made certain that the building would look the same on the outside whether the doors and windows were open or closed. To put it another way, this is a house with a moveable outer wall, instead of doors and windows. Behind these louvres are glass doors and *shoji* screens so, in effect, there are three layers of moveable walls. By manipulating these layers, the interior lighting and ventilation as well as the views from inside can be changed in various ways.

This layered system has since become fashionable in architecture, though contemporary materials are now being used. This is probably because the system is environmentally friendly and suggests interaction with the outside world.

I had not yet conceived the idea of modality when I was designing this house and chose to describe it as a 'game of cutting out views'. That is, I wanted to point out the way in which the spatial condition could be changed with time through the variability of the moveable outer wall. I also wanted to introduce an element of playfulness in the manipulation of the architecture by the occupants. Much later, when I was designing the Sapporo Dome, it occurred to me that there was a similarity between the Dome and this villa, though the two buildings are entirely different in character. Like the Sapporo Dome, this villa is in the Japanese tradition, being equipped with devices to respond to the different seasons.

The woods surrounding this house are beautiful. In the Japanese-style room, with its slightly high vantage point, one can enjoy the way in which the trees change with the seasons, and listen to birdsong. The 'aesthetic of reduction', that is, the idea that beauty emerges from the extreme reduction in scale of a large building, is a Japanese tradition. Through the reduction in architectural scale, one's eyes become drawn, not to architecture, but to nature. This aesthetic was expounded by the critic and essayist Kamo no Chomei about 1,000 years ago. When designing the villa, this idea was very much on my mind. *HH*

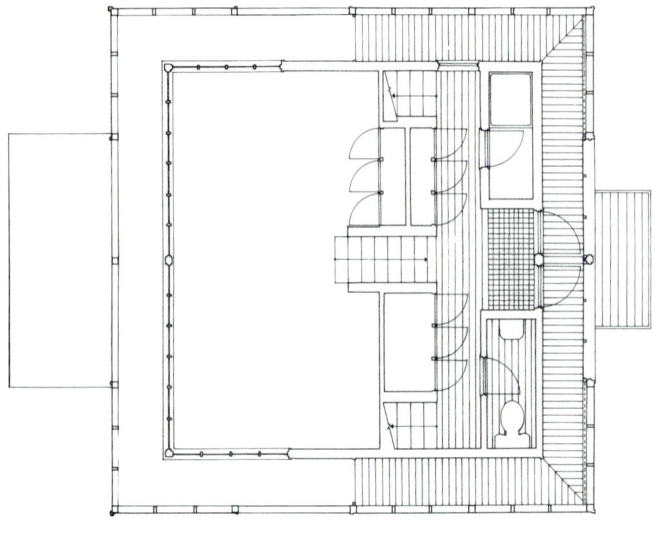

Mezzanine level plan

Interior on the second floor

Opposite: General view from the southwest

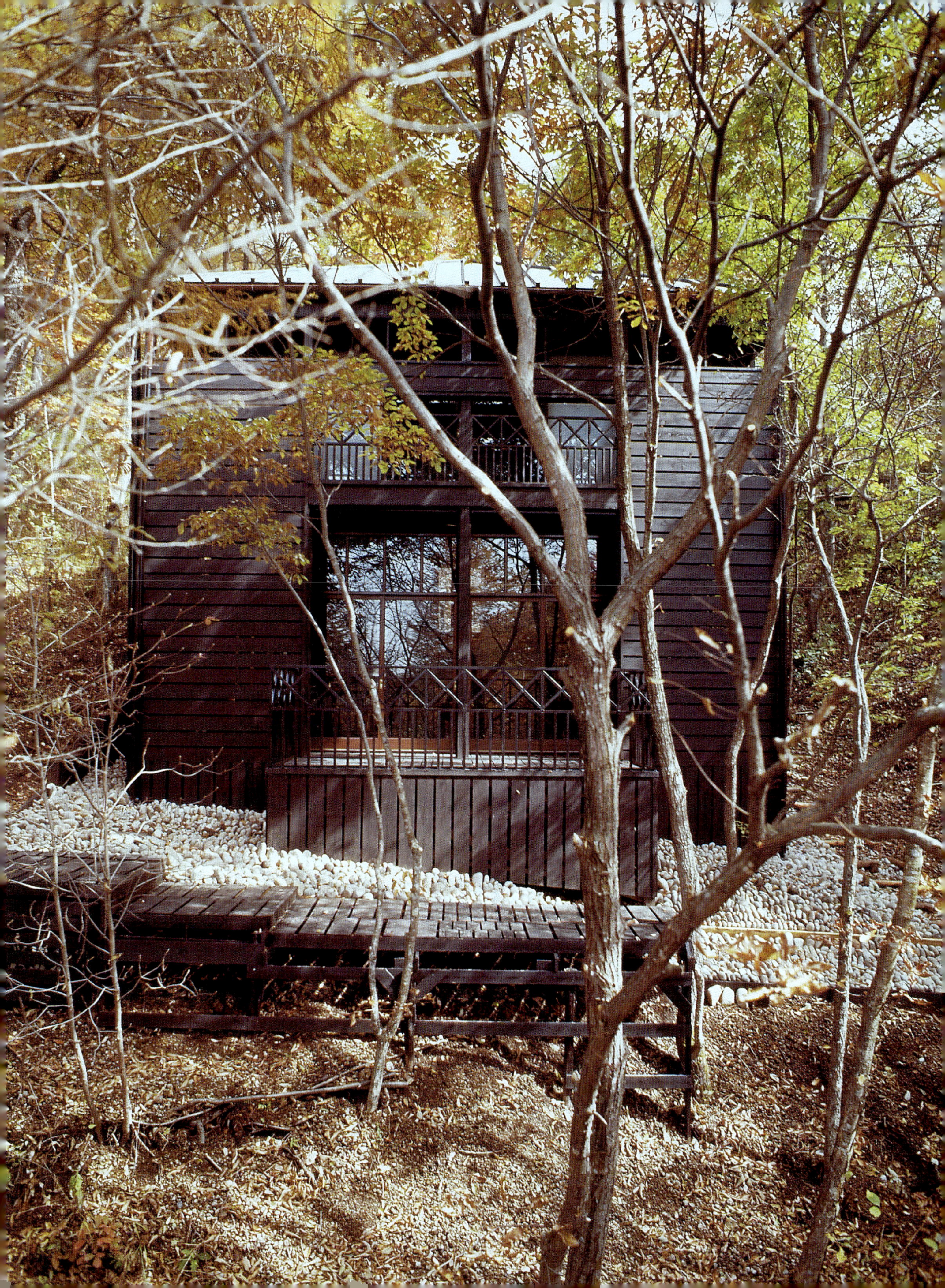

Hiroshi Hara

General view from the south

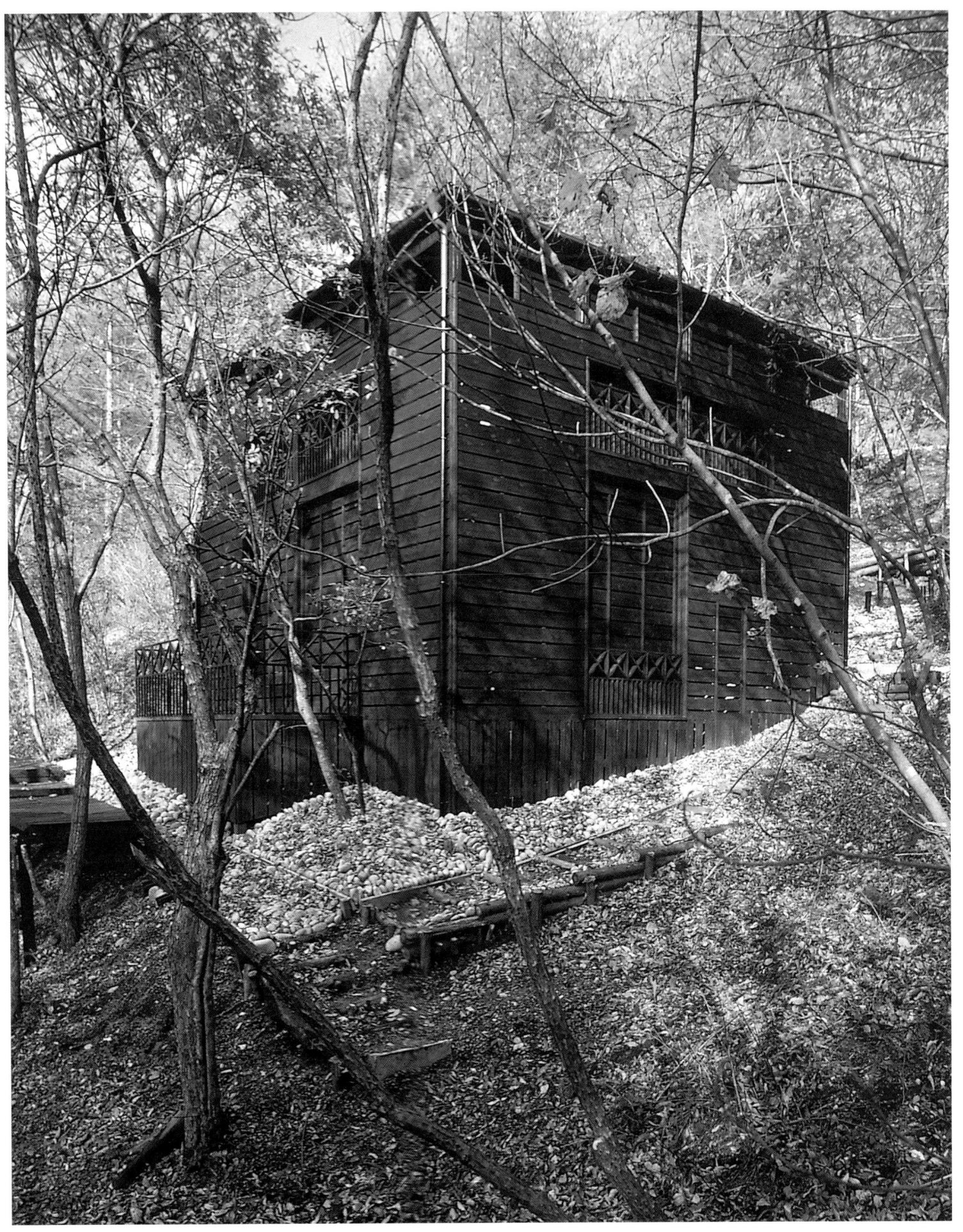

General view from the west

Niramu House, Ichinomiya, Chiba Prefecture, 1978

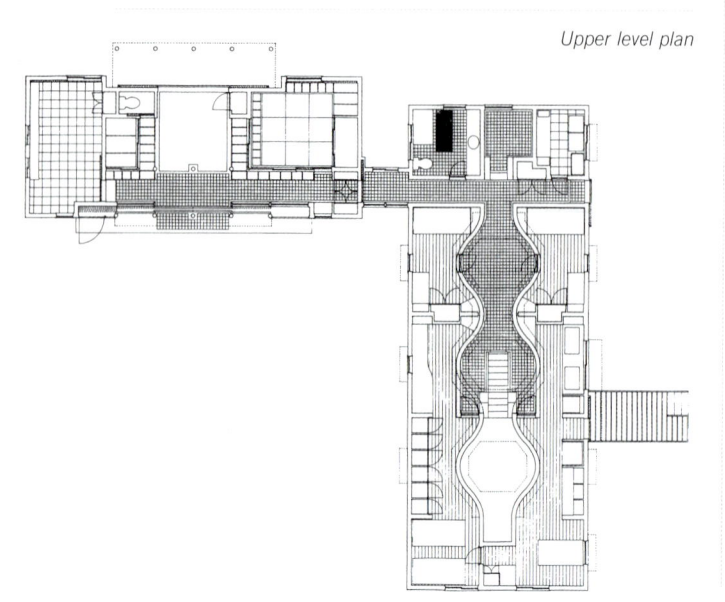

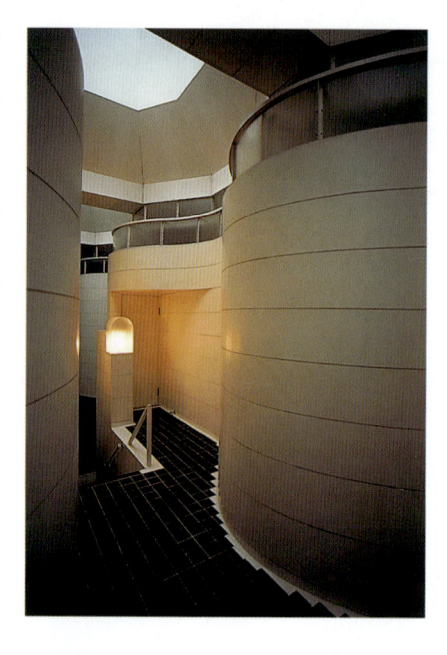

Interior; Exterior

Upper level plan

The idea of creating a dynamic 'inner core' in the middle of the house occurred to me when I was designing the Reflection Houses. This residence, otherwise based on the same formal principle as other works in the series, is the result. I considered various construction methods in order to achieve the undulating effect of the walls and eventually discovered that curved surfaces could be produced by attaching two layers of ordinary plywood sheets to studs. I used this technique in a number of subsequent works; for example, the cylindrical space of the music room in Ose Middle School.

I have great respect for the masters of modern architecture, but I also admire the architects of Futurism and Expressionism, who attempted to give expression in architecture to dynamism, which they saw as the essence of nature. They believed that architecture was a symbol of the universe; that is, their lofty aim was to represent through architecture a concept of space. The modern concept of space was undoubtedly the 'homogeneous' one of Mies van der Rohe. It was the architectural expression of a spatial concept in which humanity had believed since Descartes, and it will be some time before the architectural expression of the next spatial concept emerges. In this sense, we architects today are merely in a transitional period.

I do not believe that the seeds of the next spatial concept are in the dynamism suggested by Futurism or Expressionism. However, curved surfaces that evoke wave-motion or flow have a dramatic geometry that is different in character from a compositional architecture based on the combination of planes and solids. Of course, static compositions are themselves not without drama. There are quiet dramas as well as tumultuous dramas. Will the next spatial concept be characterised by tumult or quiet?

Such were my thoughts as I created these curved surfaces. Acrylic windows with a somewhat complex form are situated between the curved walls and the ceiling. The same detail was used in the second roof of the Hara House. Sinuosity is not synonymous with the city, but it is certainly a part of urbanism. I wanted to embed the street of such an abstract city in a small house. *HH*

Opposite and overleaf: Upward view of the hall

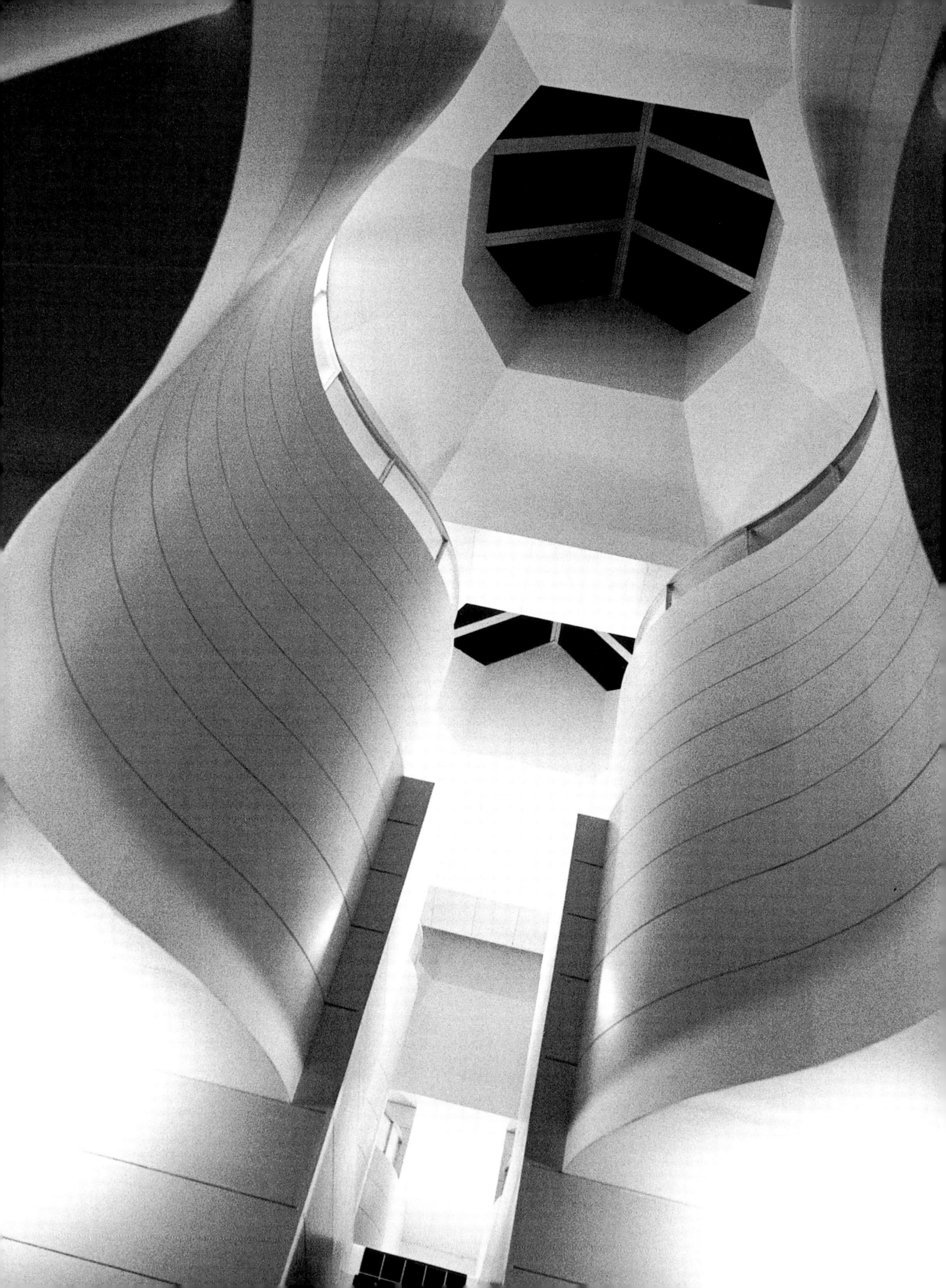

Hiroshi Hara

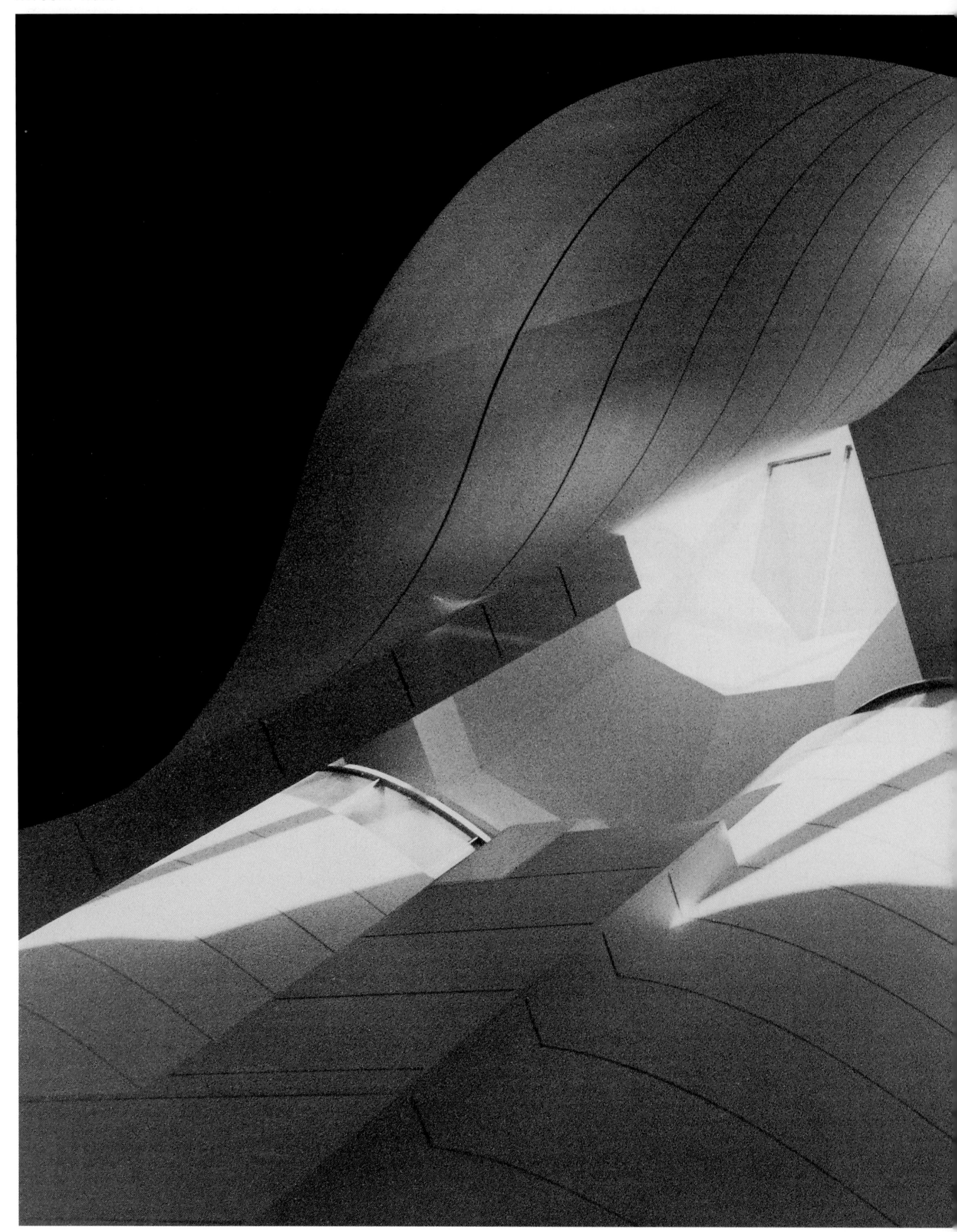

Niramu House

Niramu House: upward view of the hall

Villa *Yume-butai*, or the Stage of Dreams, Ito, Shizuoka Prefecture, 1982

This villa is based on the traditional Japanese architectural style of *butai-zukuri*, used in temples and houses. In Japan, structures built on cliffs are commonplace. In *butai-zukuri*, light wooden buildings organised according to a terrace-like structure are supported by pillars on the cliffside. The term *butai* (stage) refers to the fact that these terraces function as stages in some Buddhist temples.

The site of the villa is a cliff that faces towards Mount Fuji. The *butai-zukuri*-style terrace affords a view on to this typically Japanese landscape. Because the scenery was so picturesque, reminiscent of *ukiyo-e* paintings, the owner named the villa *Yume-butai* (the Stage of Dreams).

In contrast to the Reflection Houses, which are relatively closed to the surrounding urban environment, this villa is open to nature. The curved veneer surfaces use the same construction method as Niramu House. While Niramu House was clearly symmetrical, however, this villa consists of only one half of the symmetry. The structure is surrounded by glass surfaces. At night, these reflect the curved walls, generating an imaginary symmetry.

The idea of architecture opened up by the power of scenery played a decisive role in my building design in subsequent works; for example, the Tasaki Museum, Yamato International and the Iida Art Museum. These buildings relate strongly to the topography of their sites. This means that their placing must be exact, pointing in a specific direction. I call this *directionality*. Topography-related directionality acutely opposes the *in*-directionality of the homogeneous space. HH

North view

Upward view of the living room

Upward view towards the children's rooms from the living room

2. Multilayered Space / Self-Similarity
by Hiroshi Hara

The best example of a multilayered structure is provided by the Earth's strata, which present a record of geological history. During the 1960s this metaphor was developed by philosophers in various different ways. Today, I use the term to suggest a stack of diagrams, each drawn on a transparent plane, as often seen on computer displays. As opposed to heavy, opaque phenomena such as the Earth's strata, the multilayered structure first came to me as something light and transparent.

An example of this is a house that is light inside, so that when seen from the outside – as in the case of my own house (1974) – trees reflected on glass surfaces are overlaid on the interior view. I applied this to the Tasaki Museum (1986) and the Kenju Park 'Forest House' (1987). In these works, I thought of the architecture as consisting of two layers: the inside and the outside. Many complicated phenomena take place on both layers, but can be synthesised into one simple layer using the method of overlay. The concept of the multilayered structure is therefore a way of simplifying complex phenomena. If applied to a map, the map would become an overlay of contour lines, showing the topography along with various signs. In the case of a weather map, the pressure field is added as another layer. These architectural and cartographic examples show that the multilayered structure is a common method of understanding space, both horizontally and vertically.

In the Josai Elementary School, Okinawa (1987), I used a horizontal and a vertical overlay. Horizontally, layers of roofs remind the viewer of traditional Okinawan architecture. Vertically, the roof, ceiling and floor suggest the movement of air. Of course, the horizontal and the vertical already form the basis of architecture, in terms of plans and elevation drawings.

I have employed horizontal overlay in various architectural and urban projects. An example is the International City proposed in 1990 for a competition for a conceptual redevelopment of Montreal. This project was developed with the recognition that a set of urban elements belongs to a particular phenomenon and it does so according to a certain rule or co-ordinate. In this way, space (or layers) can be described as the outcome of the interaction between the set of elements and their co-ordinates. In general, this rule cannot be outlined as a mathematical formula. It must be provided visually by drawings and diagrams. Today, however, computers using multilayered structures to store data can do this easily, and therefore in practice there is no problem if the rule is not in the form of a mathematical formula. In the International City project, the following six layers were proposed:

Existing structure
Urban domain structure
Urban contextual modifier structure
Topological structure
Environmental structure
Urban activity and urban space structure

In the case of the International City overlay, the rules consist of the following:

Co-ordinate
Accommodators (urban residues)
Natural and urban topographical contours
Urban periphery
Urban strings

Urban strings are the virtual lines along which small devices are placed. These devices induce various urban activities.

The above concept is applicable as a general method of urban

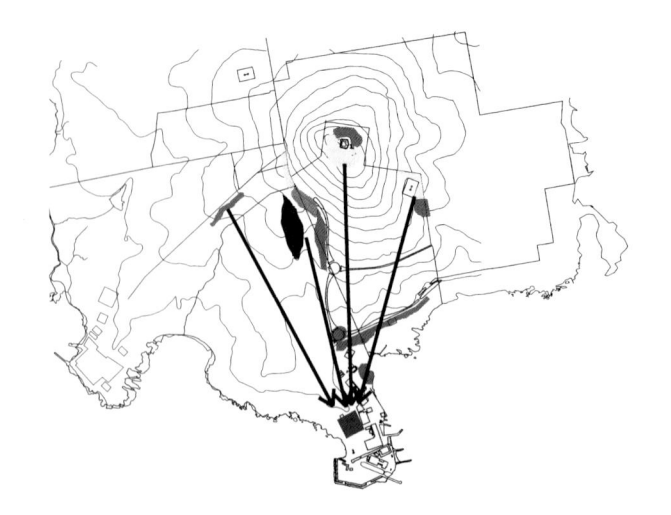

Montevideo workshop

design. In 1999, at an international workshop ('The New Space for a Global City: Cordoba'), which took place in Argentina, the theme was to develop/preserve the city's urban fabric along a river. I took this expanding site as a horizontal multilayered structure and designed it accordingly.

During March and April 2000, I participated in another workshop called 'International Seminar, Montevideo III'. The project was to redevelop a park facing Montevideo across the bay. My proposal was to renew or 'reset' the landscape and urban scenery from ten aspects, while preserving their original structure. These aspects were as follows:

   Waste-disposal
   Continuity of two beaches
   Accessibility
   Network of pedestrian paths and plazas
   Complementary vegetation
   Land movement
   Neighbourhood park
   Removal of lightning columns
   Floating wooden platform
   Small attractors

The actual 'reset' of the scenery came about through the overlaying of these ten aspects. When elaborated, each item concludes as the combination of the set of altered elements and their set of co-ordinate rules.

Thus the multilayered structure is a design method and a way to explain complex phenomena. Although each phenomenon, ie, layer, may be comprehensible, that which is the result of overlaying is usually difficult to understand. This indicates that in most architectural and urban phenomena, the multilayered structure is not linear in its composition. When a project is based on a multilayered structure, it merely means that the project analysis and design strategy are made clearer. But such a process can only be considered separately from the efficacy of the overall outcome or effect. We must always keep this limitation in mind.

The multilayered structure also handles vertical phenomena. In a traditional Japanese house, the architecture begins with a *tatami* room. Next to this is a narrow terrace (*engawa*). Then come the eaves (*hisashi*), and the balcony (*rodai*). These all extend out to the garden. At the end of the garden is a hedge, which forms a boundary with the garden next door. Seen above the hedge is the neighbouring roof, then the trees, mountains and sky. This is a standardised notion of Japanese scenery, behind which is a spatial concept based on a multilayered structure. However, in the modern city, this kind of scenery is difficult to construct. In the small Kudo Villa I attempted it by creating three layers of outer boundaries, *shoji*, sliding glass door and louvre door (*shitomi-do*). These layers frame the scenery seen from the inside.

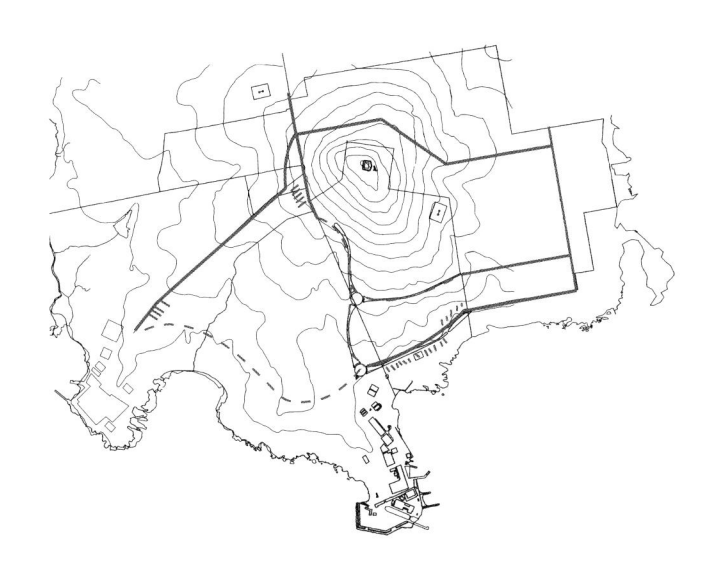

In 1984, at an exhibition that took place in Graz, Austria, I presented my first work based on the multilayered structure, although I was not yet satisfied with the results. Called *Modal Space of Consciousness,* or, informally, *Robot Silhouette*, it was an installation model of architecture and the city. A new version of *Robot Silhouette* was exhibited at the Walker Art Center, Minneapolis. Sixteen standing laminated acrylic plates and three fan-shaped plates, each set atop a pole, were carved with patterns in order to actualise the concept of overlay. These plates were illuminated via various computer-controlled light sources, which altered from time to time. Thus the multilayered device imitates transitions caused by sunlight. Though an artwork in its own right, the device is also a study model for recognising the fundamental propositions raised by the multilayered structure.

We can programme the illumination patterns of each layer of the device, but cannot easily predict its visual effect as an overlaid whole. In order to obtain an ordered pattern, we would have to adjust the device on the spot. This implies that just as in the horizontal version, we can control each layer of the vertical multilayered structure but not the final outcome. Only after some experience can we estimate the entire image. Thus we know that the correlation between the part and the whole is empirically established through a certain channel of aesthetics.

Urban design based on a horizontal overlay can also be considered from the perspective of the part and the body. Multilayered structure is a way of explaining the act of designing architecture and the city. The fundamental question of *why* we can design architecture and the city has not been explicitly thought about. Multilayered structure provides a clue for answering this question, and *Robot Silhouette* is a model for such an investigation.

Today, no computer architect yet exists, but it is true that overlay operations are best done by computers. The computer-software architect is therefore expected to utilise multilayered structure as a basic tool. In order to realise this, an architect's logic and imagination must be carefully examined and transformed into a software program. We have repetitively built architecture and cities, and I suggest that it is now time to reconsider the fundamentals of these activities.

Self-similarity is a concept that became known through the fractal theory of Benoît Mandelbrot. The basic idea is to take a certain shape and repeat it endlessly, enlarging it each time. This process, referred to as the 'nest structure', is an overlay combined with a variation in scale. If considered generally, ie, not mathematically, it would be a drama within a drama, a novel within a novel, and of course, architecture within architecture. It can thus also be seen in terms of the process of 'embedding'. Therefore embedding and overlay have some elements in common. In the Tasaki Museum (1986), I used the straight expression of architecture within architecture. 'Future in Furniture' (1992), on the other hand, proposed for an international furniture design competition, has a future city embedded in a cubic chamber.

Yamato International (1987) is not an overlay of transparent planes, but twelve elevational layers, partly chipped off. Normally a building is designed on the basis of the human activities that will take place on the plan. In this building, however, elevation planning came before horizontal planning. The complex correlation between the layers generated various spaces, and this resulted in a contemporary village within a building. Ideas of the 'corresponsive' are at the basis of my design concepts.

Yamato International vividly reflects changes in the sky. Aluminium panels and glass surfaces form a micro-topography on the elevation, and the building delicately responds to the transition of light. The elevation changes from time to time, and at sunset the entire facade glows as if the building were on fire.

The facade, therefore, becomes a field of reflecting light, where various light 'events' take place. Such events overlay each other as they would in outdoor scenery and, as a result, the

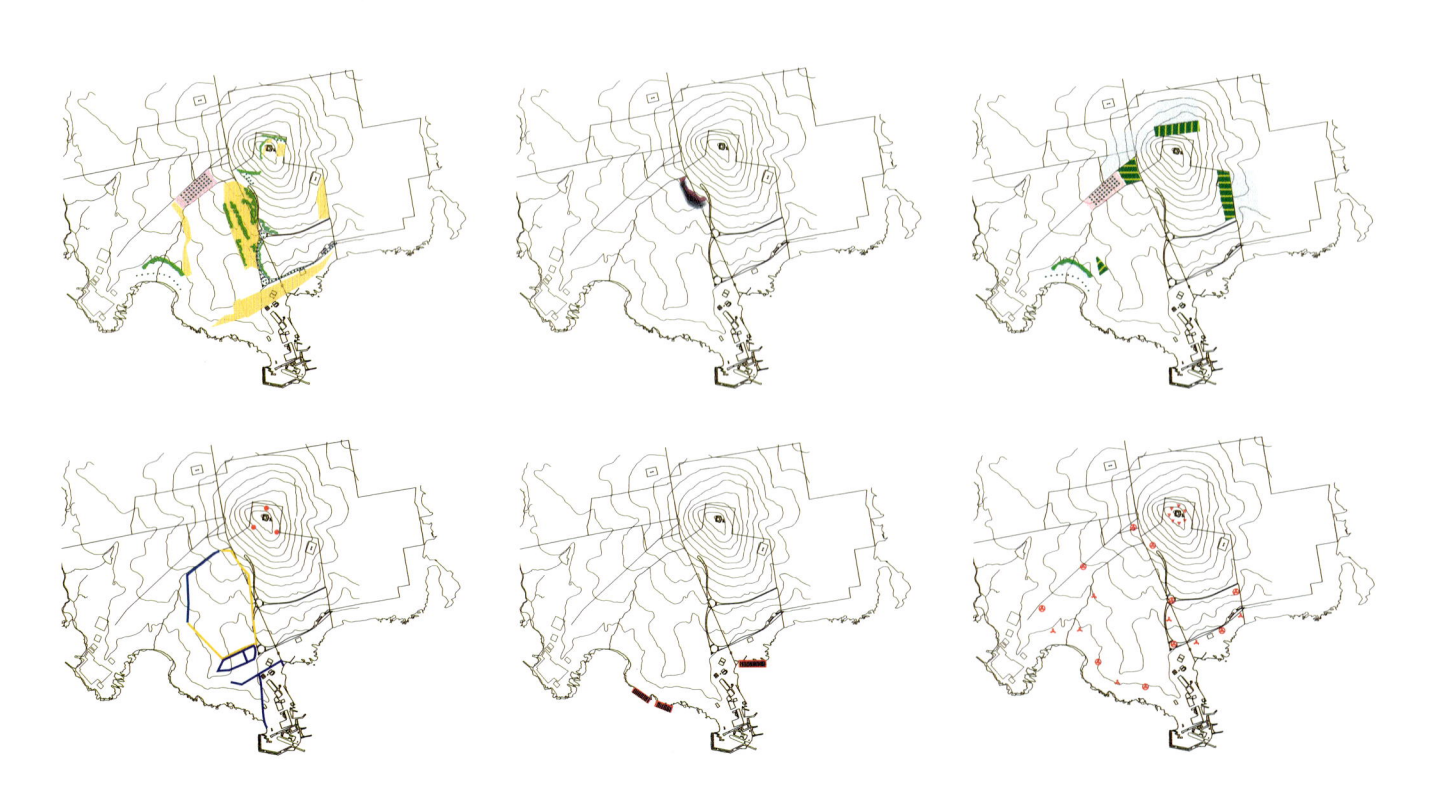

Multilayered Space / Self-Similarity

multilayeredness found in many villages is realised. I call these chronological changes of facades 'modality', based on the notion that one should never see the same architecture twice. I can anticipate and design its local effects but the whole facade is an unknown quantity until the building is finished. It is no coincidence that I wrote my book *Learning from Villages* when Yamato International was finished, since I felt that in this building I had successfully realised what I had understood from my research into the villages of the world.

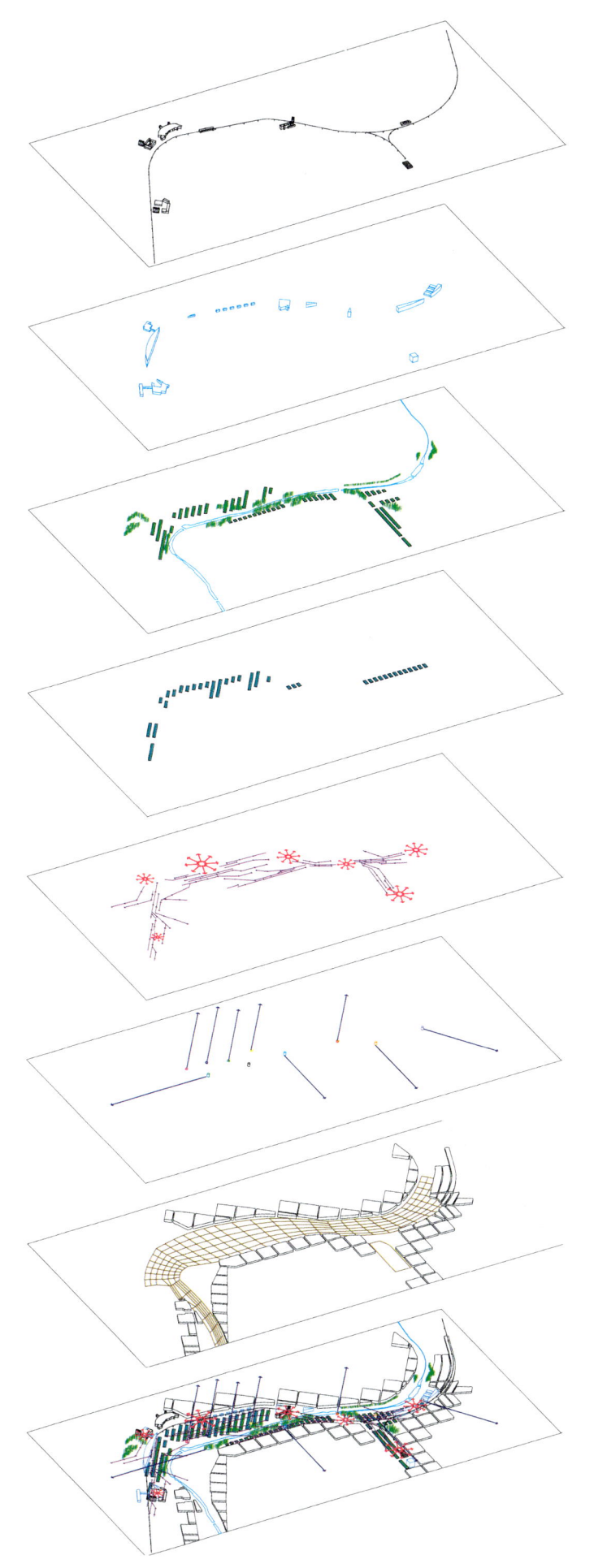

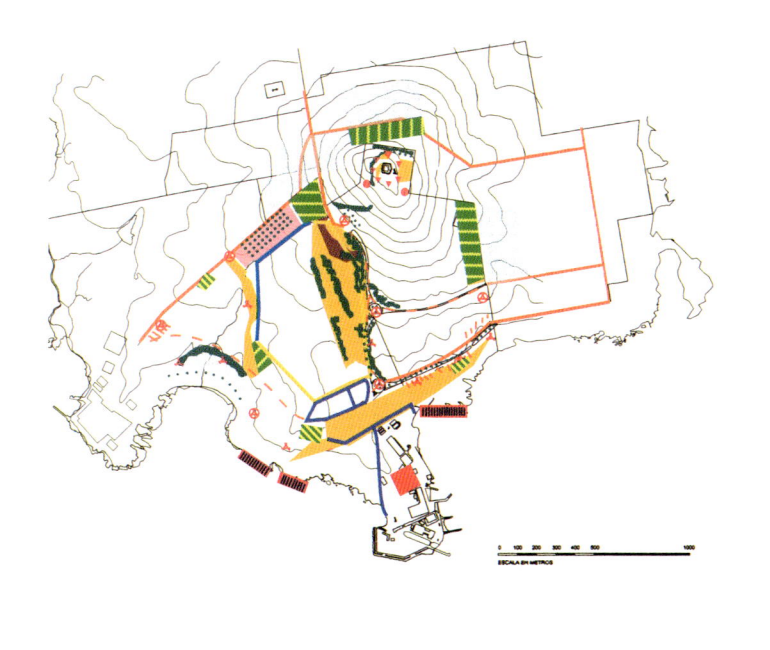

Montevideo workshop

Cordoba workshop

59

Josei Primary School, Naha, Okinawa, 1987

Okinawa consists of a group of islands at the southern end of Japan. Naha, where this primary school is located, is the central city. The site is adjacent to an historic park. Although Japan is a small country, it is extended in the north–south direction and therefore geographic conditions are diverse, and each region has its own distinctive character. Okinawa has a particularly rich culture as a result of its warm climate and unique historical background. Although most of the varied and beautiful villages that once existed in Japan have disappeared as a result of rapid modernisation, a number of traditional communities still exist on the islands of Okinawa. The area around the site was destroyed during the Second World War, but using as reference photographs of the area before the war, and villages surviving in other places, I designed this primary school in the form of a traditional Okinawan village.

'Open schools' – primary schools that ideally have no walls and provide an unstructured education had recently been introduced, and it was clear that they had both merits and demerits. In this case, however, the open-school form was chosen mainly for environmental reasons. Okinawa, with its warm, virtually winter-free climate, made ventilation a matter of the highest priority in the design of an interior climate. Even if partitions were provided, they were likely to be opened while the classrooms were in use. Taking a hint from villages in Okinawa, where each house is organised around a courtyard, I arranged wall-less classrooms around courtyards. The result was a school with a continuous roof, articulated over each classroom.

This is the only time I have designed a building that is similar in appearance to the villages of its region. My architectural approach has usually involved a contemporary interpretation of what I have learned from villages around the world. I believe that what we can learn from villages are not forms but principles and abstract aesthetics. I made an exception in this case because the school is adjacent to a historic park, and to a symbolic gate. Okinawa was the scene of fierce fighting during the Second World War and was under the control of the United States long afterwards. During that time, the rest of Japan was unable to extend help to its inhabitants. I designed this school with a painful awareness of the past that. *HH*

Courtyard

Opposite: Close-up view of the roofscape from the northeast

Josei Primary School

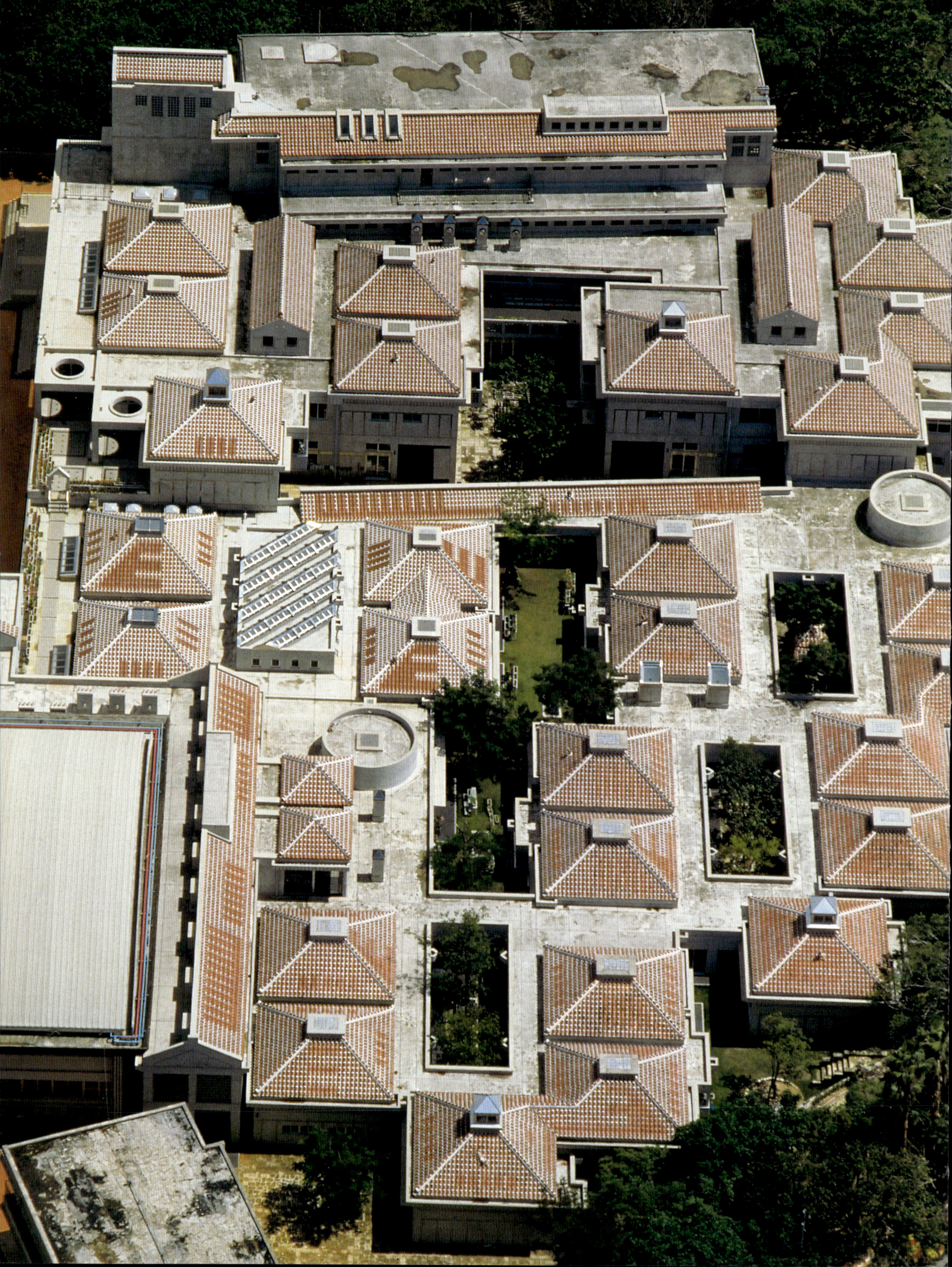

Josei Primary School

View of the roofs and the courtyard from the northeast

Opposite: Bird's-eye view from the southwest

Hiroshi Hara

Upward view of a classroom

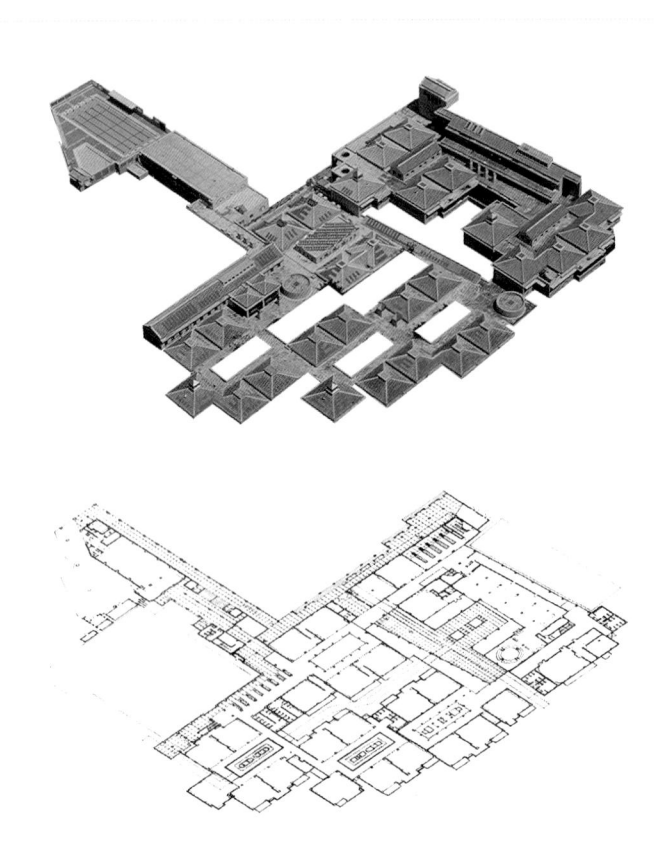

Josei Primary School

The communal area seen from the classroom

Yamato International, Tokyo, 1987

By the time I received the commission to design this building, I had already developed several schemes for multilayered structures and spaces. A garment-manufacturer commissioned the building to accommodate its offices and workshop. The structure is about 120 metres in length and 40 metres in height.

Unlike a house or some other small-scale building, a structure of substantial size has a facade that impacts on the city. I wanted to see if I could create a facade that was spatial in character. My solution was to make this element a multilayered structure.

This facade is not an orderly composition of materials and proportions but is determined instead by the way light happens to be reflected off its surfaces. These reflections differ according to the condition of the sky and sunlight. When the sun is high in the sky, the facade is characterised more by the contrast between light and shadow than by reflections. However, as twilight approaches, the reflections seem to interact and at times envelop the entire facade. If there is a beautiful sunset, the entire facade will become a radiant golden colour. This is what I call 'mixing light' and 'designing air' on a facade. The multilayered structure of aluminium panels and glass surfaces induces these phenomena. At Yamato International it is a means of transforming the material facade into space.

Although not every phenomenon that takes place on the facade can be anticipated at the time of design, it is important to impose some order on the events that may occur, using intuition or simulations of possible modalities. When I was designing the building, I relied on memories of phenomena induced by multilayered structures in villages I had visited. The most important example was the cluster of small cities in the valley of M'zab in Algeria, which Le Corbusier is also said to have visited. Although they are made of entirely different materials, the small cities of M'zab become as phosphorescent at twilight as Yamato International.

From the first time we met, the late company president, Mr Hannya, and I enjoyed a relationship of trust built on our common Buddhist world view. He truly understood the idea that we should never see the same building twice. *HH*

Opposite: Exterior view from the northwest

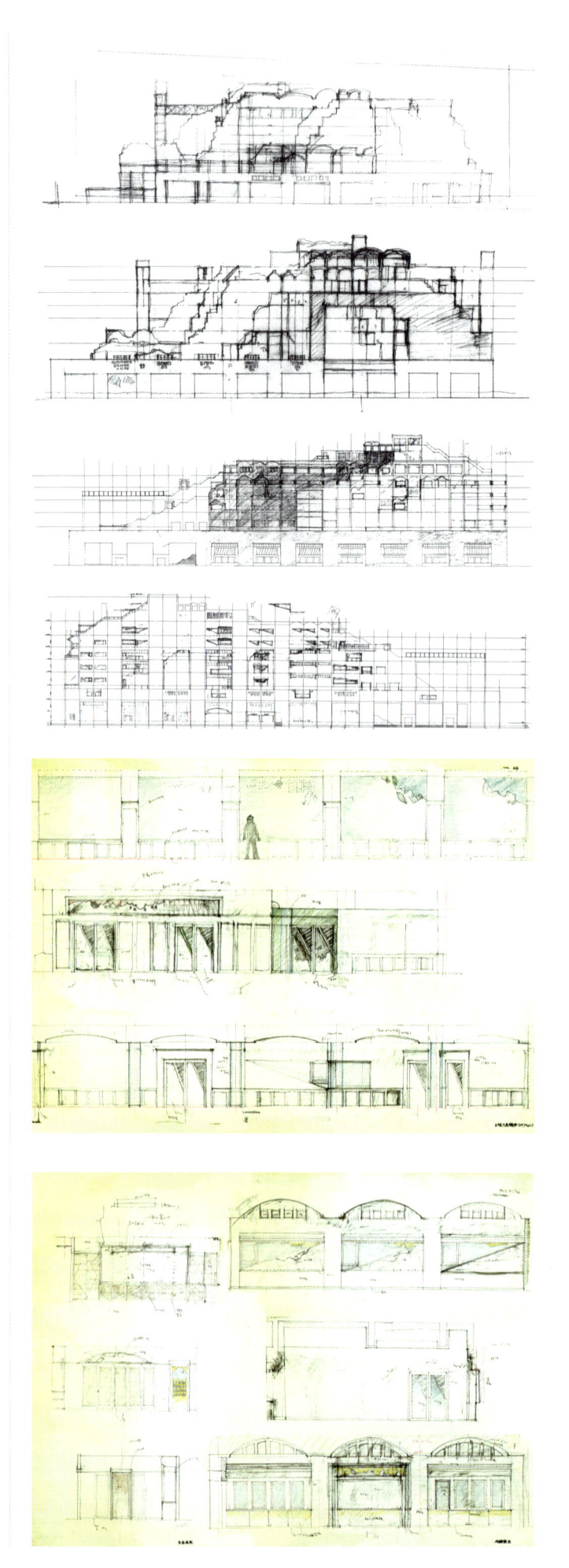

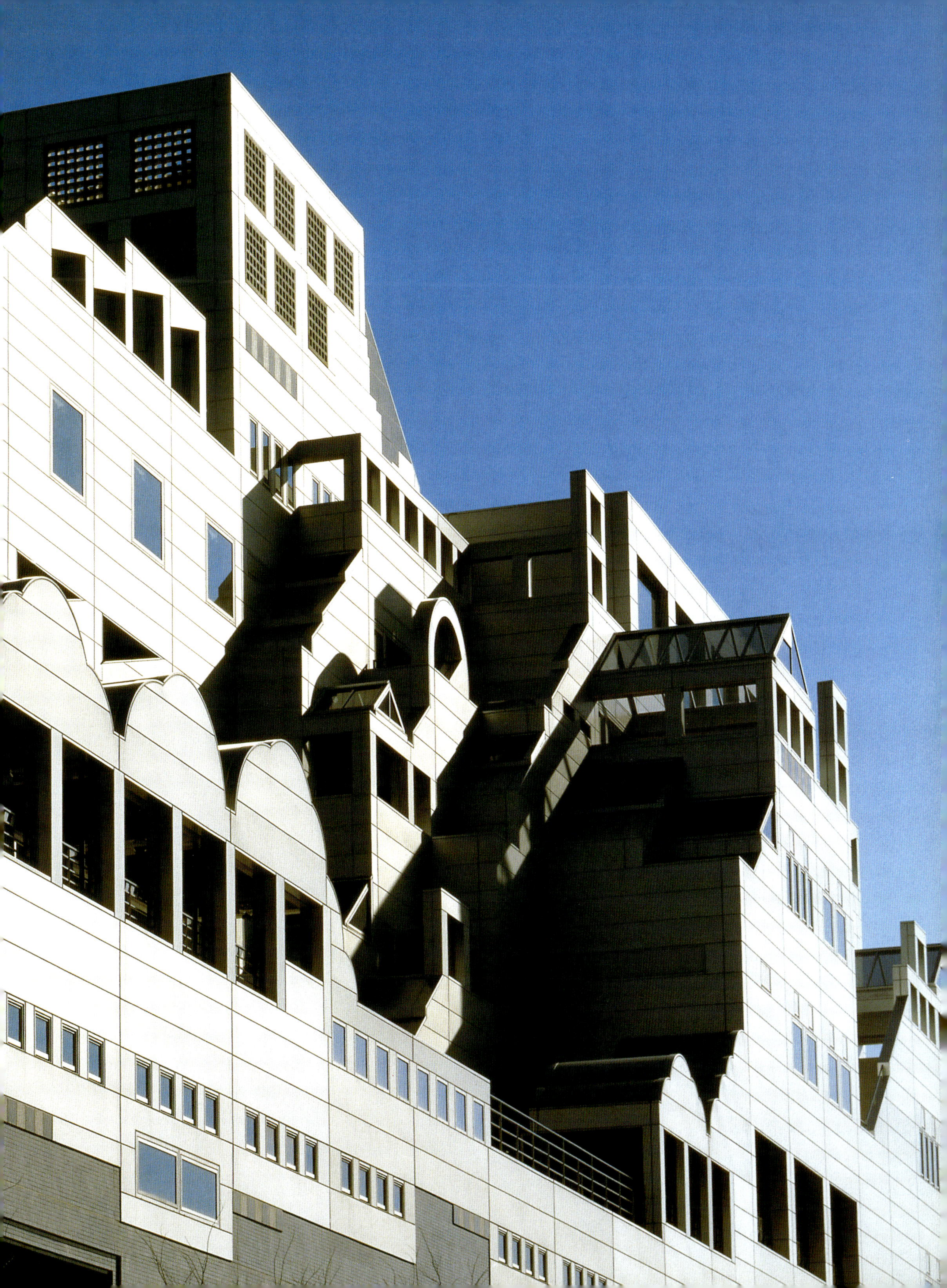

General view from the southwest

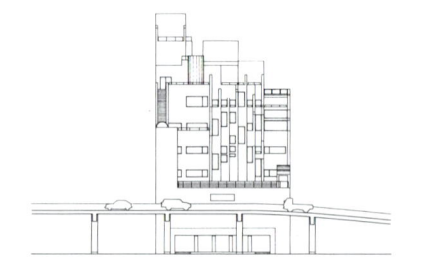

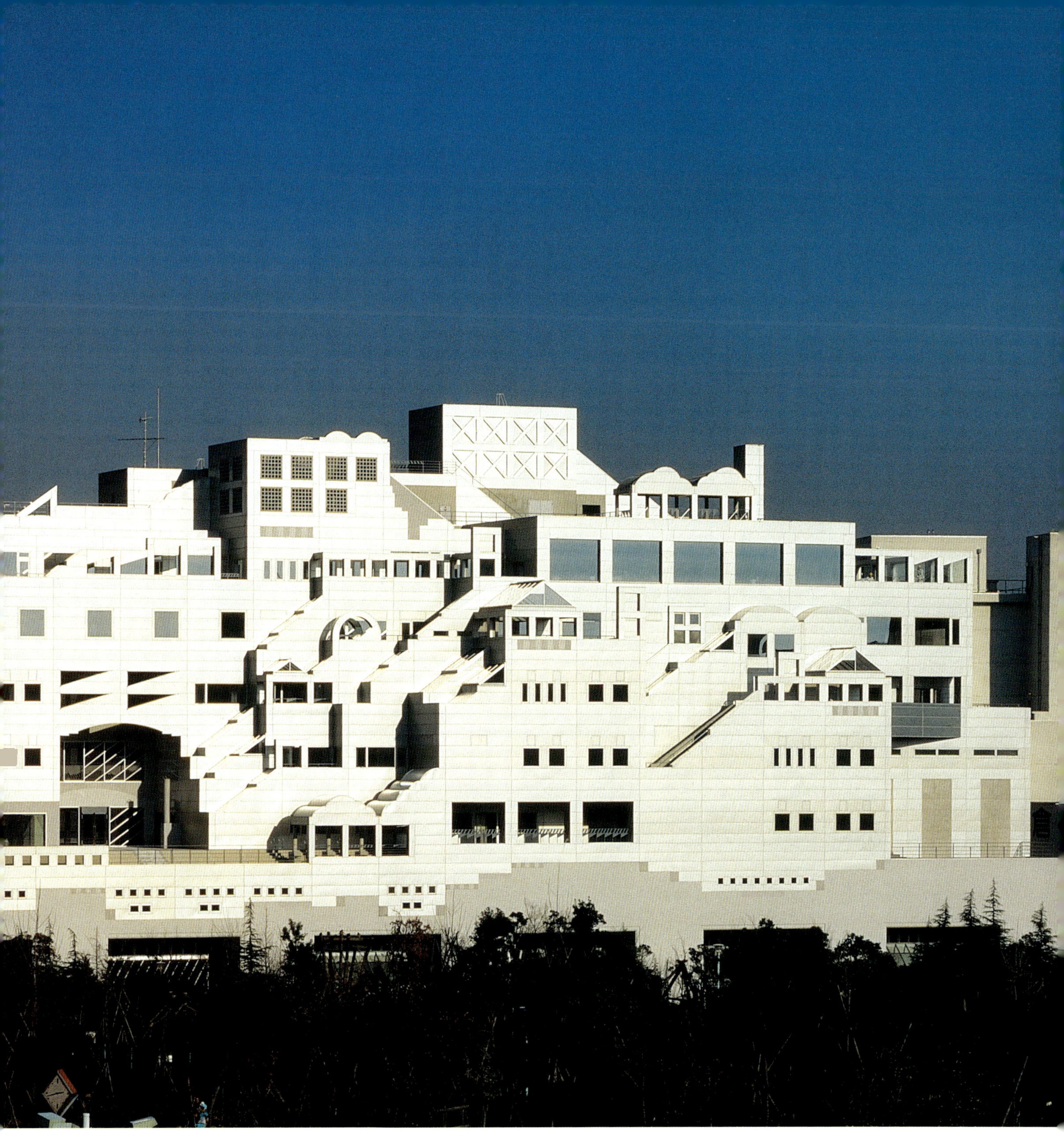

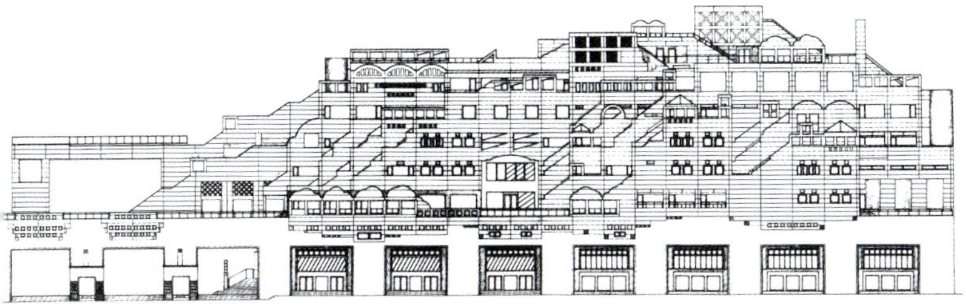

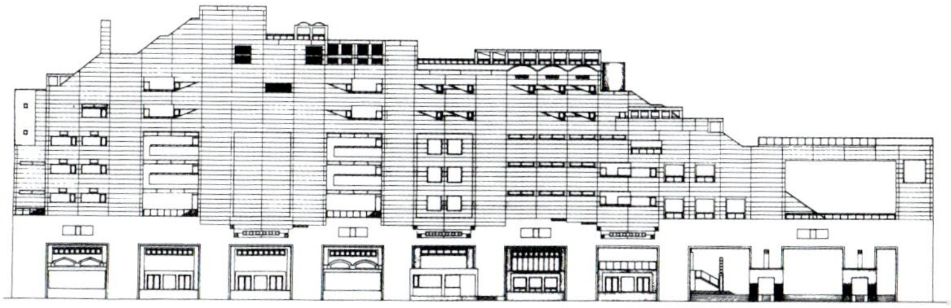

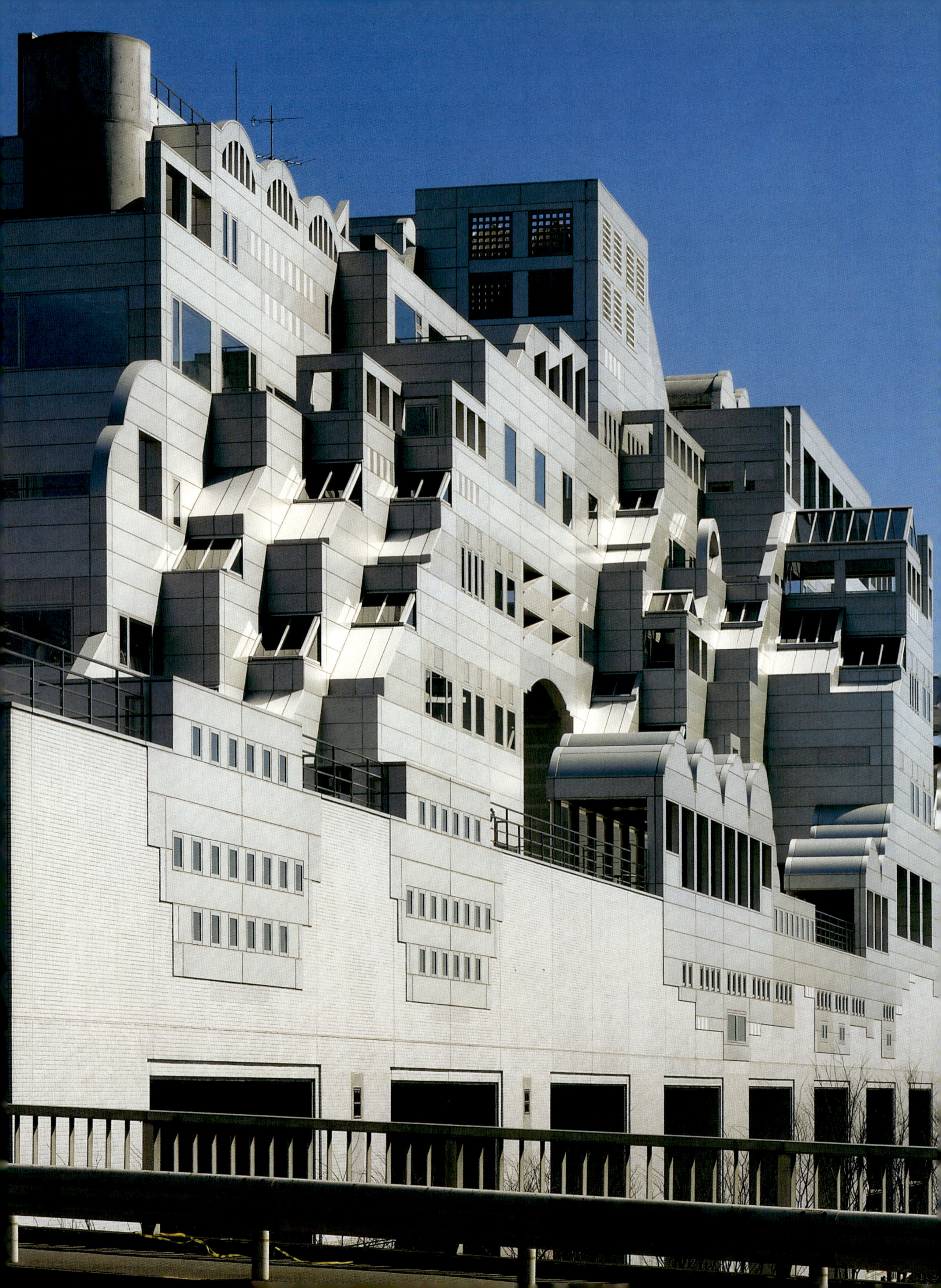

Yamato International

Close-up views from the west

Hiroshi Hara

View towards the courtyard seen from the entrance hall

View from the loading dock

Close-up view of the jagged glass between the entrance hall and the loading dock

Hiroshi Hara

Exterior view from the fourth-level terrace

Yamato International

Overleaf: West facade in the evening

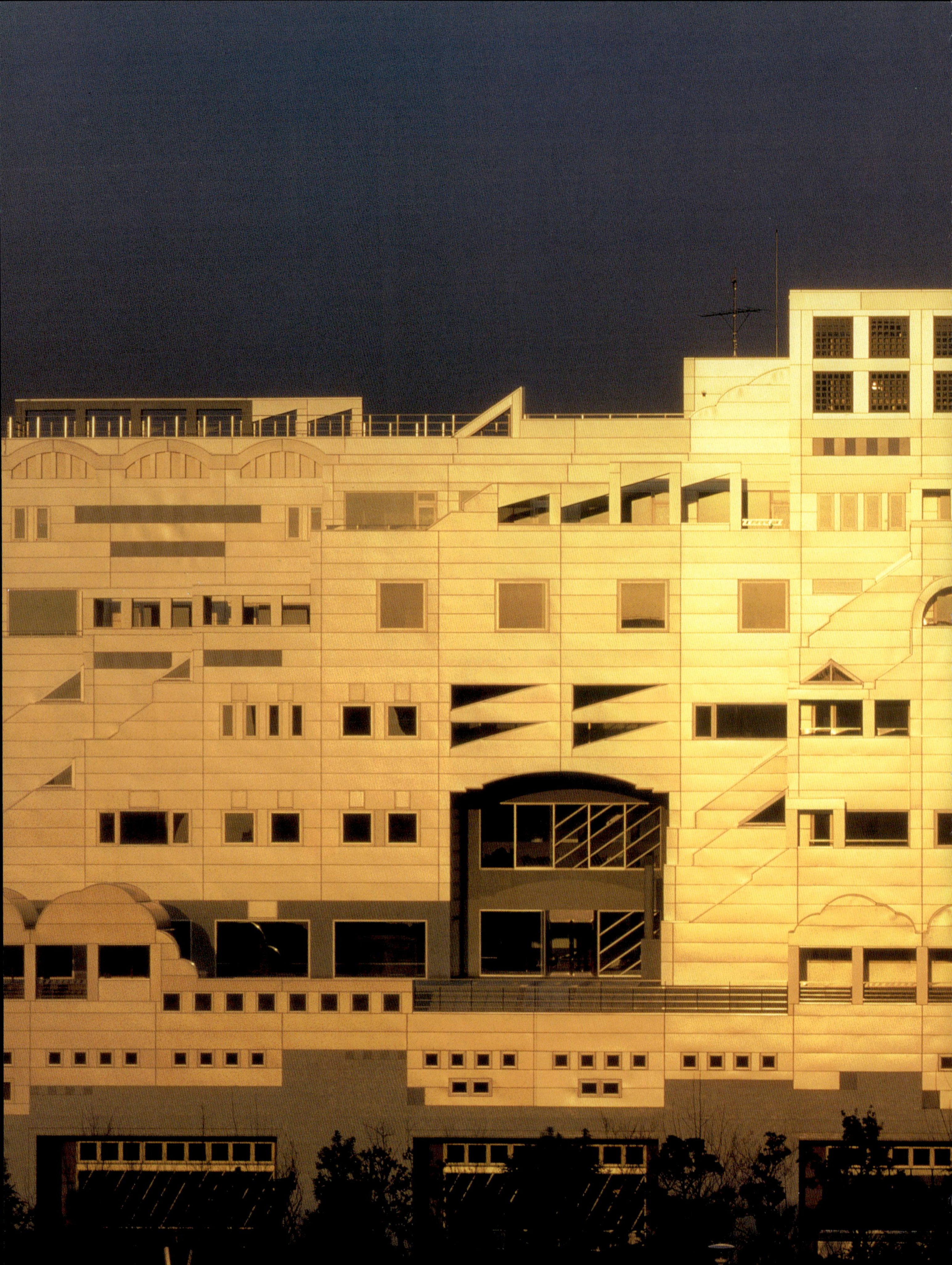

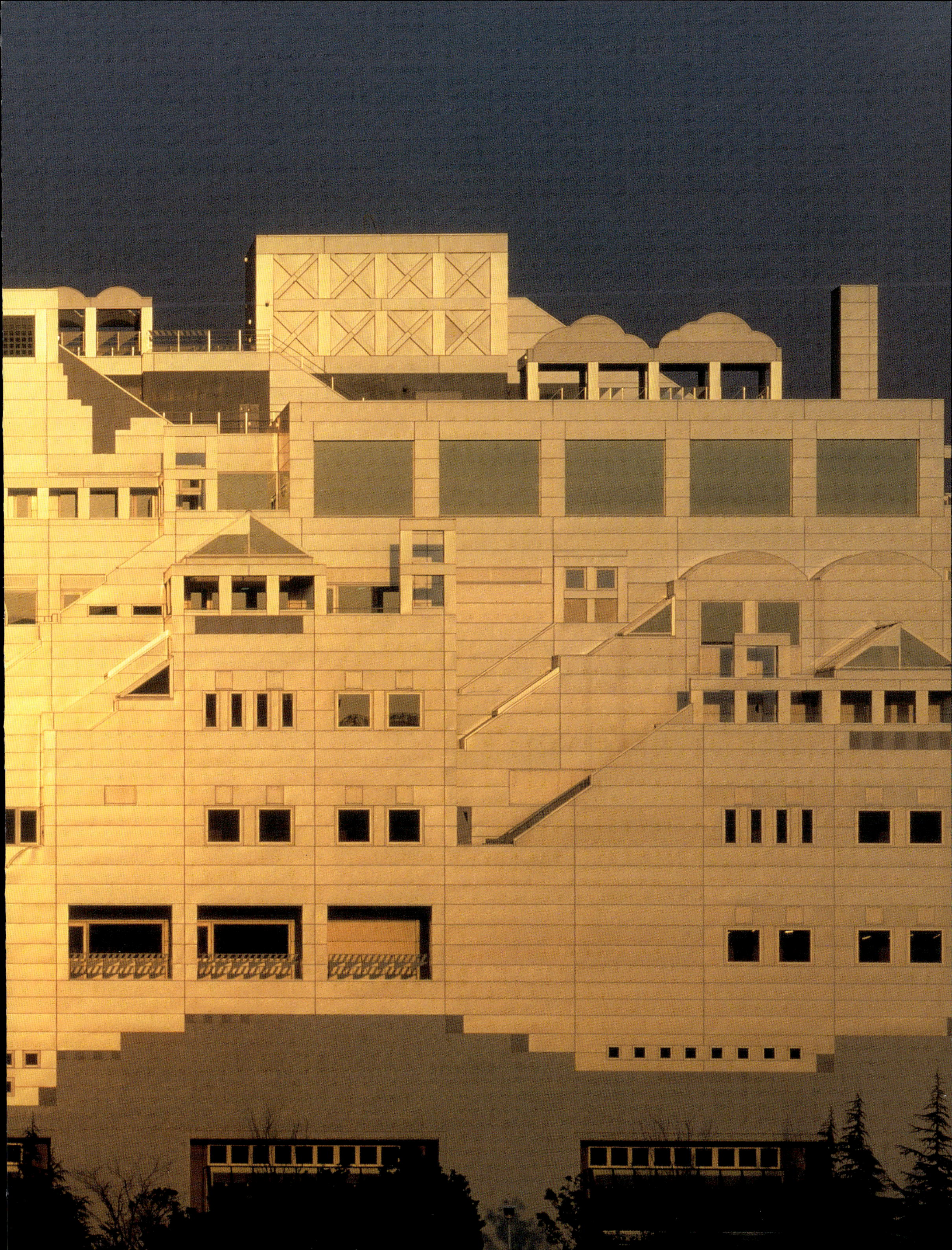

Yukian Teahouse, Ikaho, Gumma Prefecture, 1988

The teahouse is one of the most splendid building forms in traditional Japanese architecture, and many masterpieces exist in the genre. The medieval period in which the teahouse developed into its final form was aesthetically the zenith of Japanese history. Almost everything of excellence in Japanese culture with which the world is acquainted today was established in this period in the fifteenth and sixteenth centuries. This originated in the eleventh and twelfth centuries, when culture flourished in Heiankyo (present-day Kyoto), with the decision by artists to achieve Buddhist ideals through art. At the time, Buddhism made a distinction between sacred and secular ways of life. Although secular, artists aspired to live in the sacred world. Over the course of 400 years, they developed the teahouse in order to realise that aspiration

Thus it is almost meaningless for an architect today to design a teahouse. Simply put, today one can create only a teahouse-*like* work of architecture. From the time I was given this commission, therefore, I made it clear that I did not intend to create something akin to the original model. Today, it needs to be acknowledged that the tea ceremony is theatre, not reality, and that it takes place not in a teahouse, but on a stage set made to suggest one.

I decided to create a simple shelter with two Japanese-style rooms inside it; the rooms are not completely enclosed by walls or ceilings. Since a genuine teahouse is a small building with a roof and ceiling, I was satisfied that what I had created resembled a stage set.

The aesthetic of the teahouse is based on the Buddhist idea that 'being' and 'not being' have the same meaning. The forms of the teahouse completed in the medieval period were architectural expressions of 'not being' that were intended to evoke in people images of 'being'. My intention was to pay homage to the forms that artists of the past developed to transport one from the secular to the sacred realm. However, my work oscillates within a secular framework. I designed a stage set in an ornamental manner in order to indicate 'being' directly. Creating a genuine-looking teahouse might be permissible in film or theatre but not in architecture.

The aesthetic of the teahouse is both profound and far-reaching in its implications. One of its concepts is *utsushi*, which means making an exact copy and transferring it elsewhere. *Utsushi* is in fact Japanese for both 'copy' and 'transfer'. The ornamentation in this stage set of a teahouse is a collage of fragments of *scenes of teahouses* that remain in my memory. The teahouse is essentially a fragmented *utsushi*. The resulting building represents the embedding of images of the teahouse. HH

Cutaway axonometric

Interior perspective

Opposite: Corridor

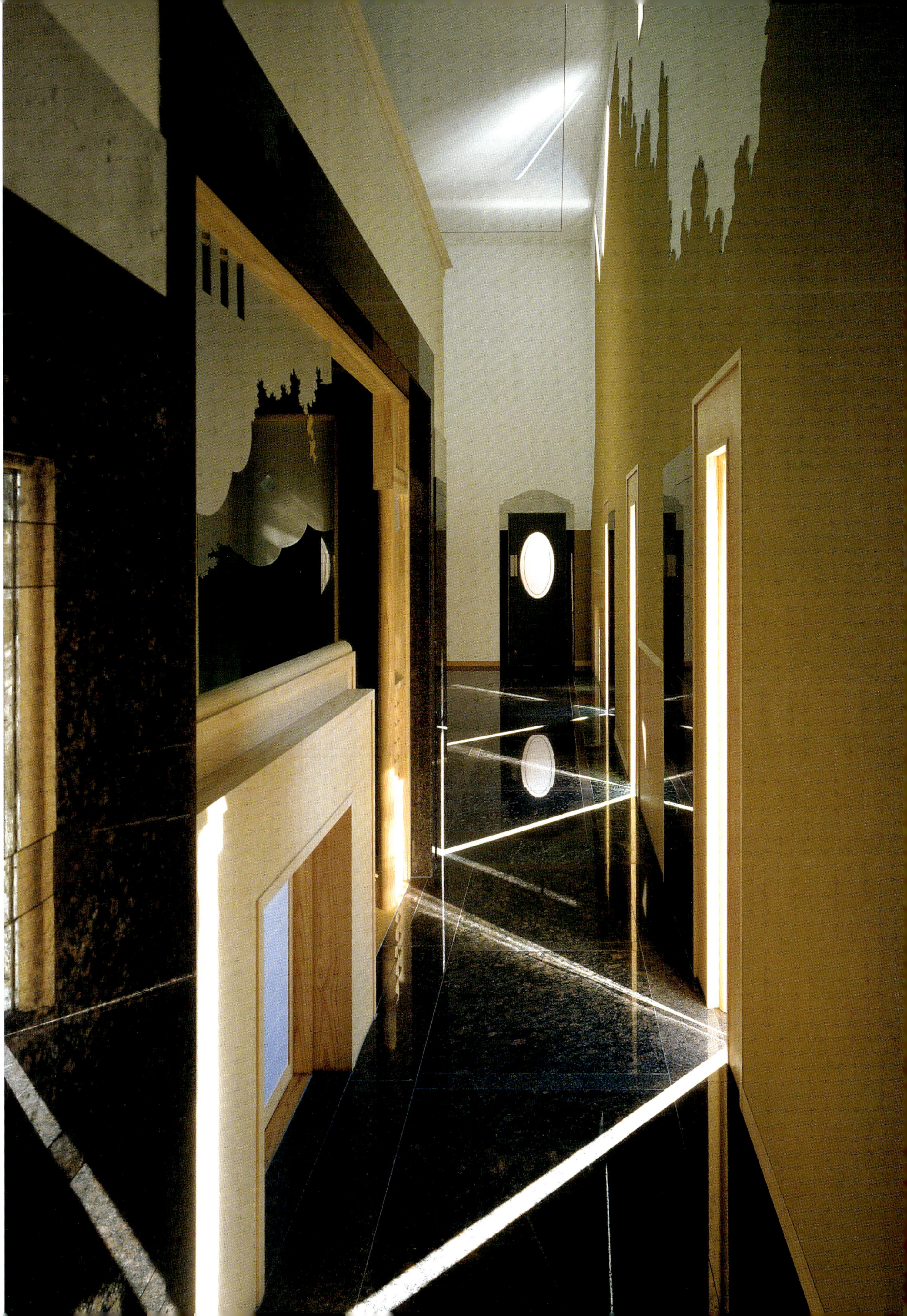

Hiroshi Hara

Yukian Teahouse

Four-and-a-half-mat room

Opposite: Corridor

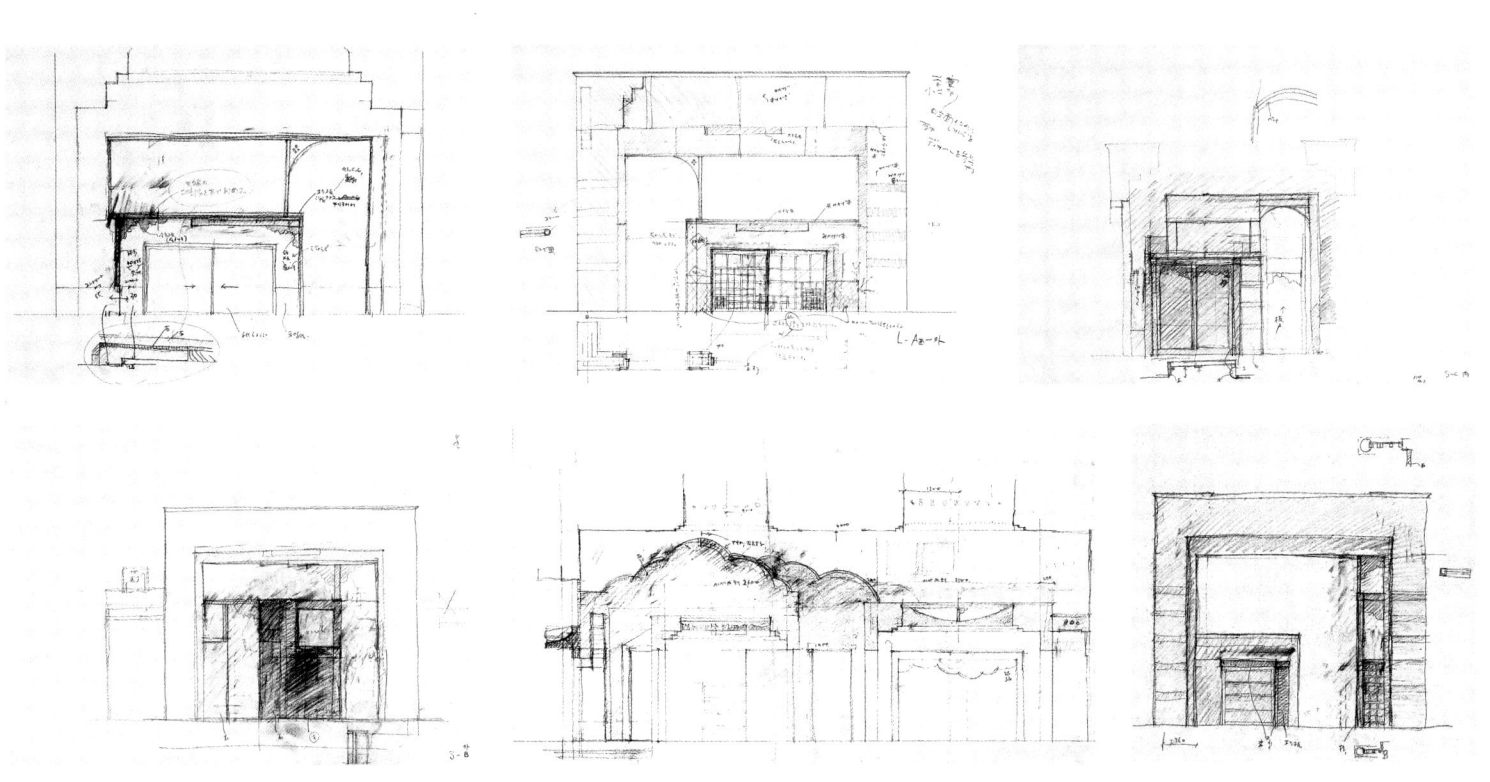

81

Hiroshi Hara

Future in Furniture
(competition entry, first prize), 1992

Furniture – the *future* can be found in furniture

These ideas were proposed for an international furniture design competition. The concept was to produce a furniture kit that embodies a miniature model of mid-air cities like the Umeda Sky Building and the 500-metre Cube. The kit is assembled within an empty building frame.

'Urbanscape Furniture' examines the possibility of a city embedded in furniture. In other words, furniture is transformed into a set of tools for the conception of future cities. Different types of furniture will generate different types of city. Firstly, we designed a lamp as a model for a new housing type; then we tried to realise a complete interior environment from such pieces. Such a *furniturised* room becomes an urbanscape, a 1:60 scale model of a floating city. This process illustrates the possibility of furnishing a complete city. Thus Urbanscape Furniture is a general concept that includes all possible types of furniture designed through similar procedures.

Shown here in detail are designs for a room unit, a balloon-shaped lighting tower, and a chair, all having various functions, dimensions, materials and methods of assembly. *HH*

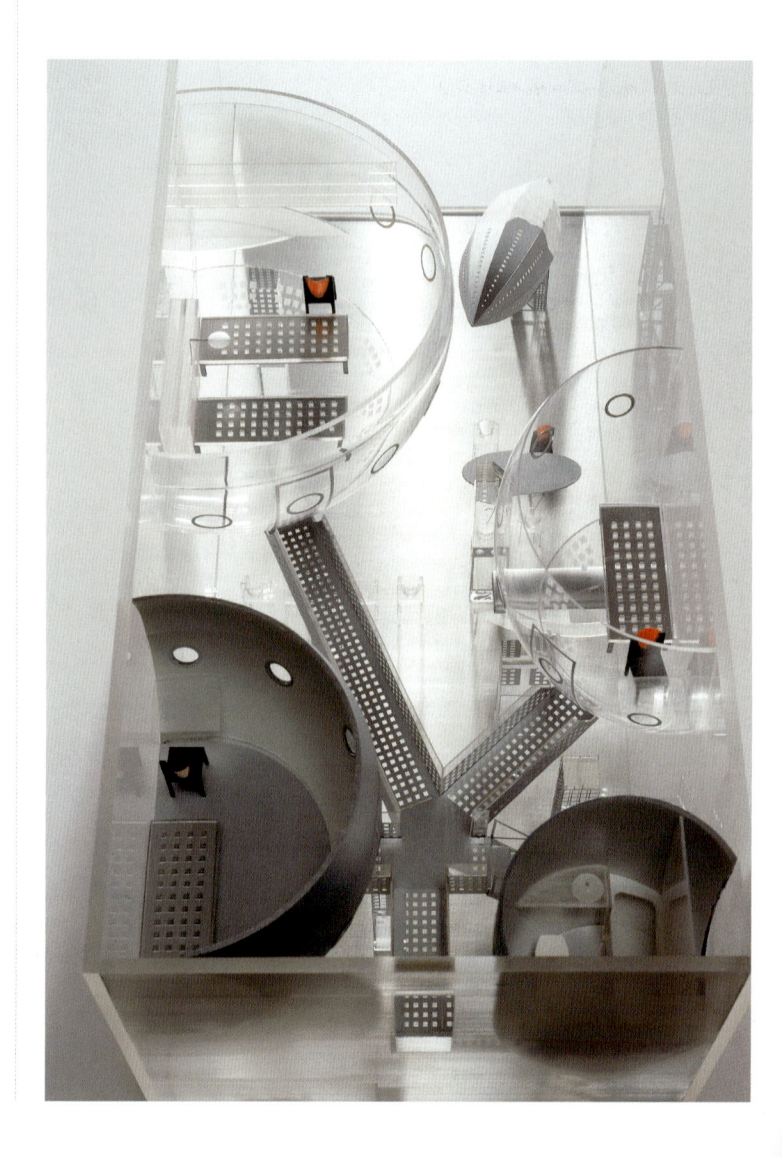

Model photo

Future in Furniture

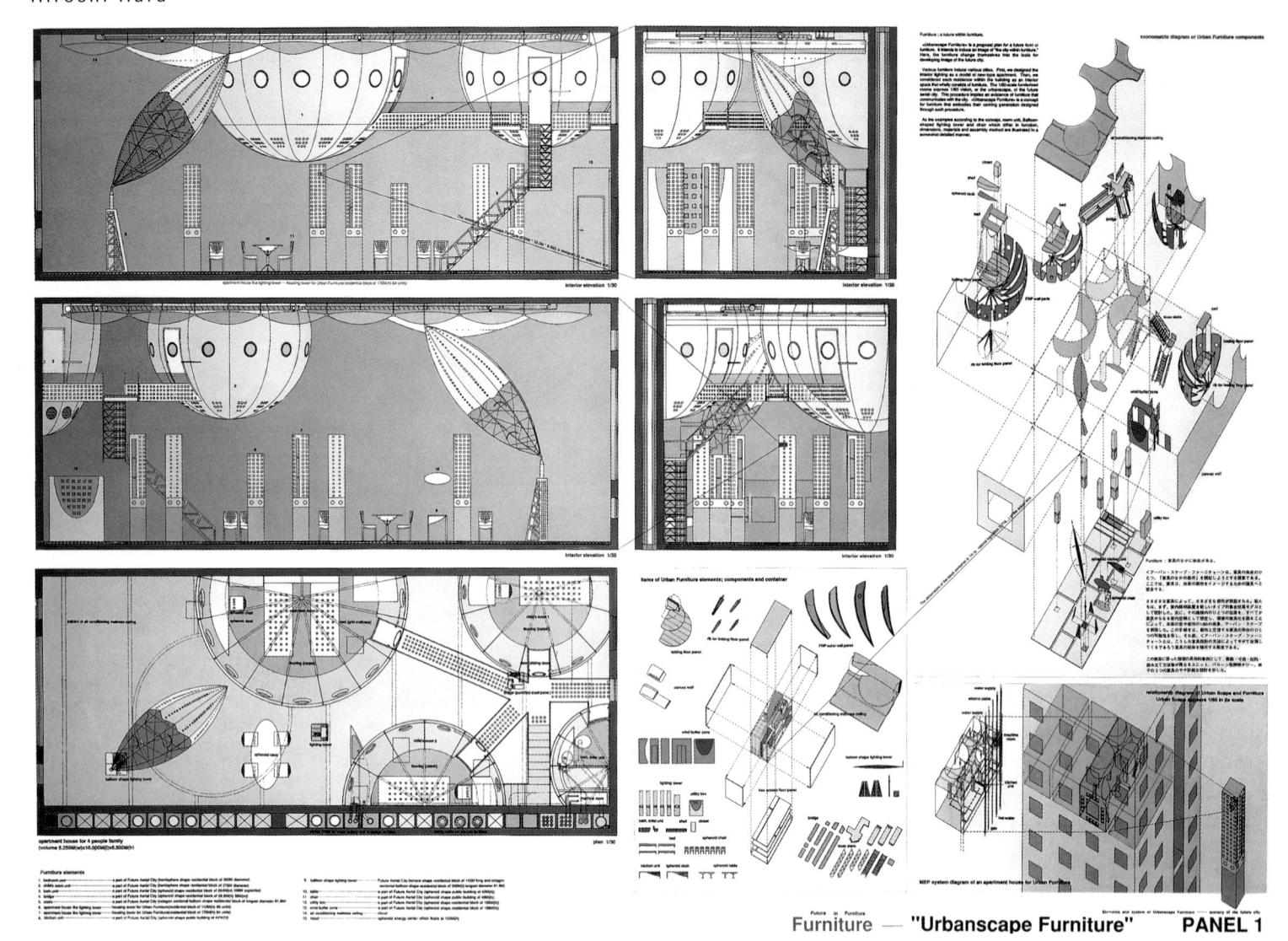

Hiroshi Hara

Furniture — "Urbanscape Furniture"   PANEL 1

Future in Furniture

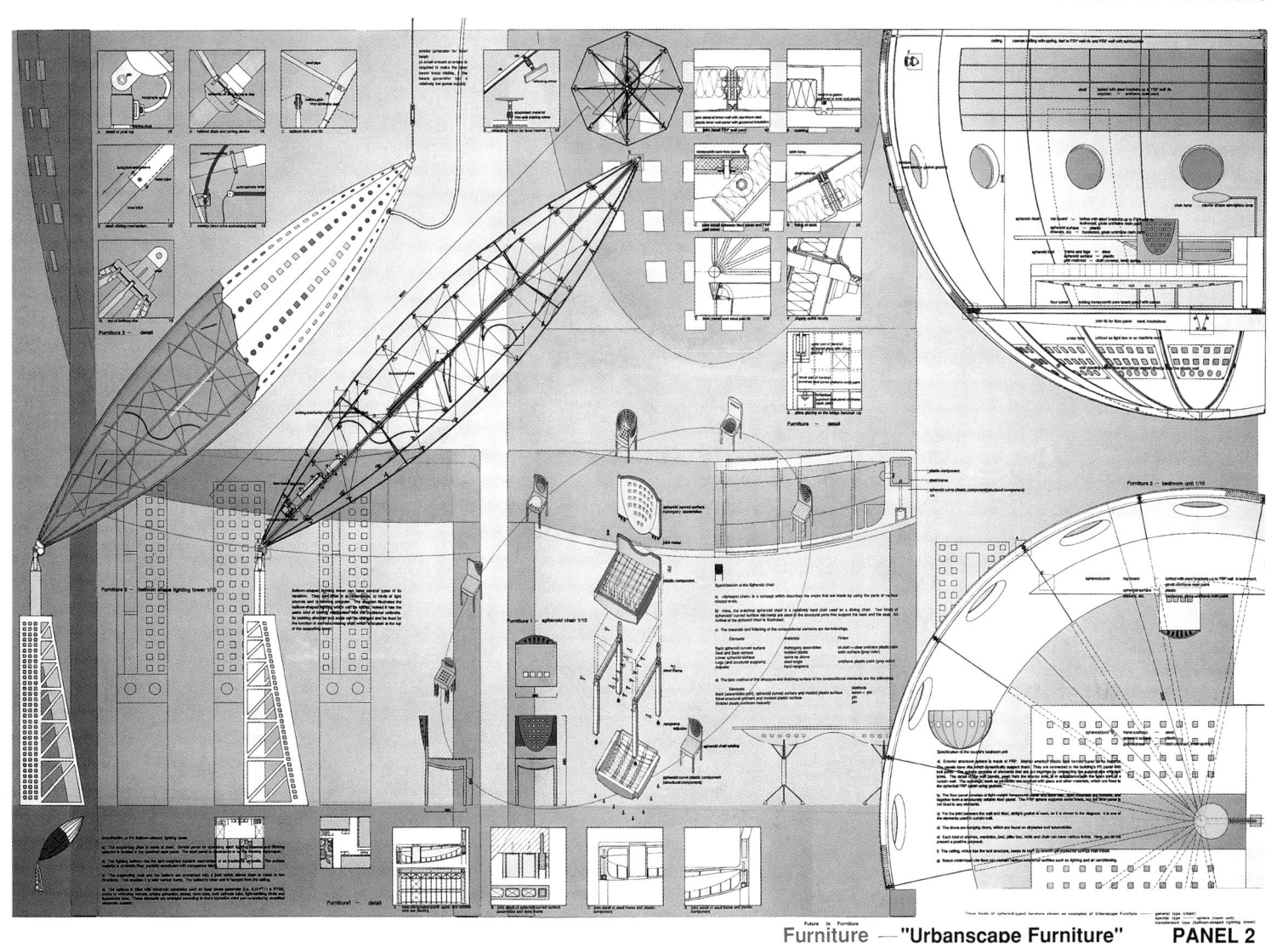

Furniture — "Urbanscape Furniture"   PANEL 2

3. Transposition / Incidental Illusion
by Hiroshi Hara

If architecture is regarded as a 'thing', then the model is for an unchanging condition; if considered as an 'event', then architecture is in constant flux.

It is almost the same thing to see architecture as an event as to understand it as space. Such an understanding also makes it possible to recognise architecture as a field. In architecture, fields appear with regard to temperature, light and air movement and they are in a state of constant change. It is clear that these conditions are not things but events. If buildings are designed close to nature then these environmental or interior climate fields should change in accordance with nature. One aspect of architectural space is the overlay of these fields. Environmental and interior climate fields can therefore be considered to have a multilayered structure.

People experiencing architectural space traverse these fields as events. I began to consider this notion of 'traversing' when I started to design large buildings in which people move from one place to another within its interior space. However, I have always been interested in the movement of people experiencing and observing architecture.

Among the fields, that of light is the most important in architecture. The design of a field can be symbolically dubbed the 'design of air'. This implies that when someone enters a building or a room, they feel that the quality of air is completely different from outside, but in fact such an experience depends on the design of light. In houses and other small buildings, I tried to design 'different air' using natural light entering from skylights and high side-lights as well as light reflected from patios. I then realised that an unavoidable element is the reflection of light on glass surfaces. I discovered that by utilising reflection, I could introduce fictional phenomena to architecture.

The transparency of the glass surface has a character that can be described as 'border but not border'. Indeed, this is a basic concept in medieval Japanese aesthetics, which can be linked to Buddhist philosophy. For instance, according to Buddhist teachings, to exist and not exist are the same, and there is no border between the two. More directly expressed, living and dying make no difference. To the dialectical Western way of thinking, this total contradiction may appear impossible to overcome, but in Eastern philosophy it can be achieved by way of nirvana. The logic that overcomes, or simply accepts, this absolute contradiction is called the logic of 'both-and'. The artists of medieval Japan tried to realise the world of nirvana through their expressions. This is conveyed by the design of tea-ceremony houses (*chashitsu*) and traditional village houses.

Reflections on a glass surface, when overlaid on to the interior

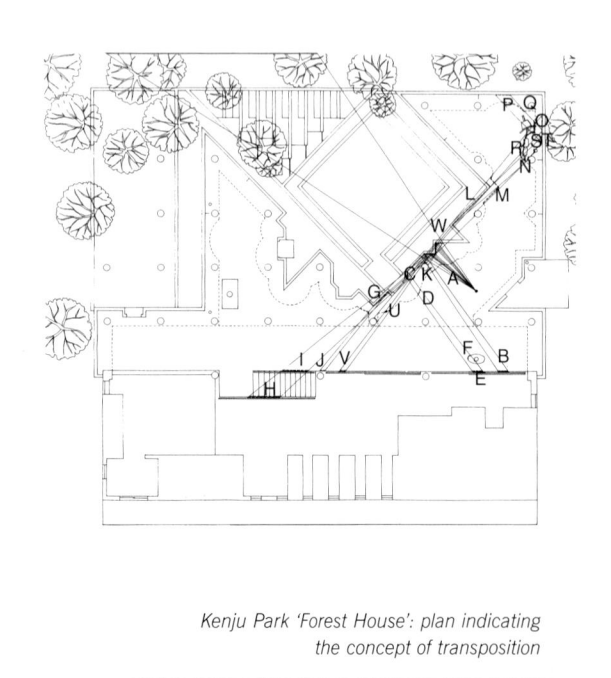

Kenju Park 'Forest House': plan indicating the concept of transposition

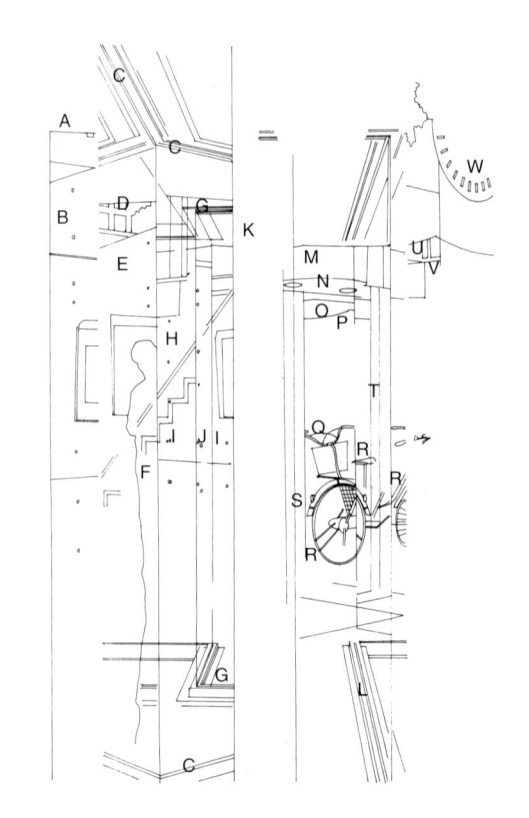

Illusion

Transposition / Incidental Illusion

view, produce a sort of collage. Borrowing from the Surrealist notion of *depaysement*, or transposition, the reflected view is considered to have been displaced from its original position and transferred on to the glass surface. This results in the 'transportation of the sky' or the 'transportation of the woods'.

As exemplified in the Tasaki Museum (1986) and Kenju Park 'Forest House' (1987), when glass surfaces are set in a zigzag configuration, they simultaneously reflect parts of the building and the surrounding environment. Depending on the observers' position, the time of day and the light condition at a particular moment, unexpected views will appear. A slight variation in the position of the spectator will make an enormous difference to the view. This difference is brought about by a careful consideration of fictional effects. For instance, in the Tasaki Museum, one person will see a fictional view of the patio surrounded by trees while another person who is standing elsewhere will see the real patio with no trees. These individual events, or tricks, can be designed by calculation; but the same can never be said of the overall effect. In the Tasaki Museum, when zigzag glass panels are set among the cloud-roofed buildings some interesting images evolve.

Following my experience of the Tasaki Museum, I wanted to place zigzag glass surfaces everywhere in the 'Forest House', which was surrounded by foliage, to make the outside and the inside of the house ambiguous. A photograph of this house might therefore appear to be a collage of the forest and the interior view.

Such transportation of nearby elements like foliage, and of more distant objects like the sky, can give a building a fictional character. However, in real urban situations, the concept of transportation is also present. Parks, for example, are composed of elements like trees and ponds transported from elsewhere. A building can also be said to have been transported – perhaps from a textbook on modern architecture – since it did not originally exist in that location. So both the city and architecture have fictionality through transportation as their basis. These were fundamentals that I wished to express in the Tasaki Museum and 'Forest House'.

TS Eliot's *The Waste Land* (1922), which was groundbreaking as the first poem to use the method of collage successfully, twice makes reference to the 'unreal city'. This phrase had a crucial effect on me in my youth. As Eliot himself has stated, it is a phrase inspired by a passage from Charles Baudelaire's *Les Fleurs du Mal*: *Fourmillante cité, cité pleine de rêves, ou le spectre en plein jour raccroche le passant*. My understanding of the phrase 'unreal city' was that it referred

Reality

Kenju Park 'Forest House': transposition photo

to some sort of a fictional city. I believed that buildings and cities should be 'unreal' and this conviction has never changed. Of course, 'unreal' is a word full of implications and possibilities.

One interpretation of 'unreal' is the notion of artificiality. It is true that any architecture is artificial, but that does not mean it is unreal or fictional. Conventional architecture is real. In terms of technology or design, completely new architecture, when it first appears, seems unreal and fictional. But when the technology becomes common and the designs have imitators, unrealness and fictionality fade gradually regardless of the building's historic value. Nevertheless, unique and aesthetically perfect buildings that have a control and balance which nobody can copy remain fictional.

Far from the academic history of a generalised Classical architecture, certain villages across the world have specific forms that fit the particular topography of their site. These unique devices make them both fictional and real. For example, there are villages in Iran that use artificial oases as their device, while in Yemen skyscrapers are employed. Lake Titikaka's village with floating islands and Iraq's village where each house is built on an artificial island should only exist in the realm of fantasy, but they are real.

In general, architectural form does not accommodate itself to a single answer; there are various possibilities. It is not difficult, therefore, to make a design bizarre or strange, but such buildings are often uninteresting. They are neither unreal nor fictional. Architecture has an unreal atmosphere and fictional character only when it employs a unique and intelligent device.

Large buildings heavily based on technological requirements, such as Osaka's connected skyscraper, the Umeda Sky Building, the Kyoto Station, or Sapporo Dome would lack attraction without devices. When designing these large-scale urban complexes, I bore in mind the small fictional buildings I had designed earlier, and the particular devices I had employed in them. Such concepts are dealt with in each chapter of this book.

The changes that take place in these buildings, for example those that correspond to the transition of nature as well as double images and illusional scenes or visual tricks, come into being only momentarily or through slight differences in the observer's position. For instance, in my own house, complete symmetry in the interior space, including sunlight, is only realised at a certain moment. This is also true of Yamato International's sunset-reflecting facade, which is totally dependent on the condition of the sky. However, these phenomena do not happen by mere chance. As long as architecture employs the devices that induce such unique phenomena, it is corresponsive rather than stochastic.

Architecture based on illusion does not question one's physical being but rather one's consciousness. In contrast to the Corbusian physical architecture of the early part of the twentieth century, in the latter half, many architects intentionally aimed at the architecture of

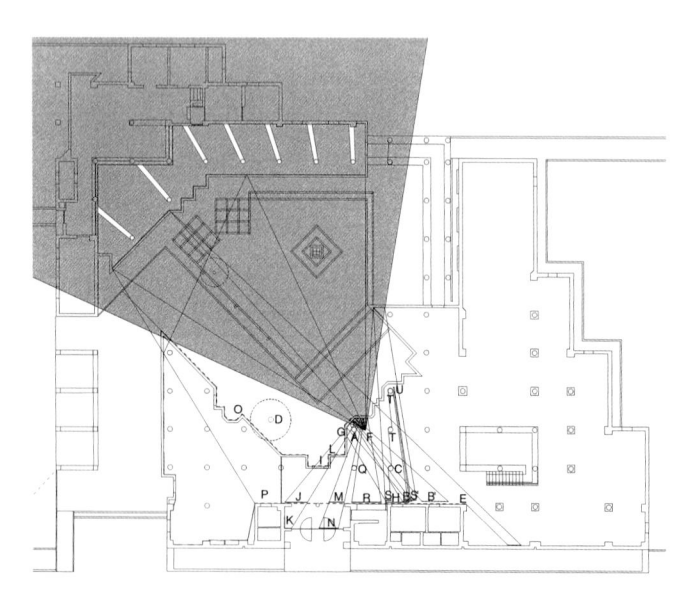

Tasaki Museum of Art: plan indicating the concept of 'transposition'

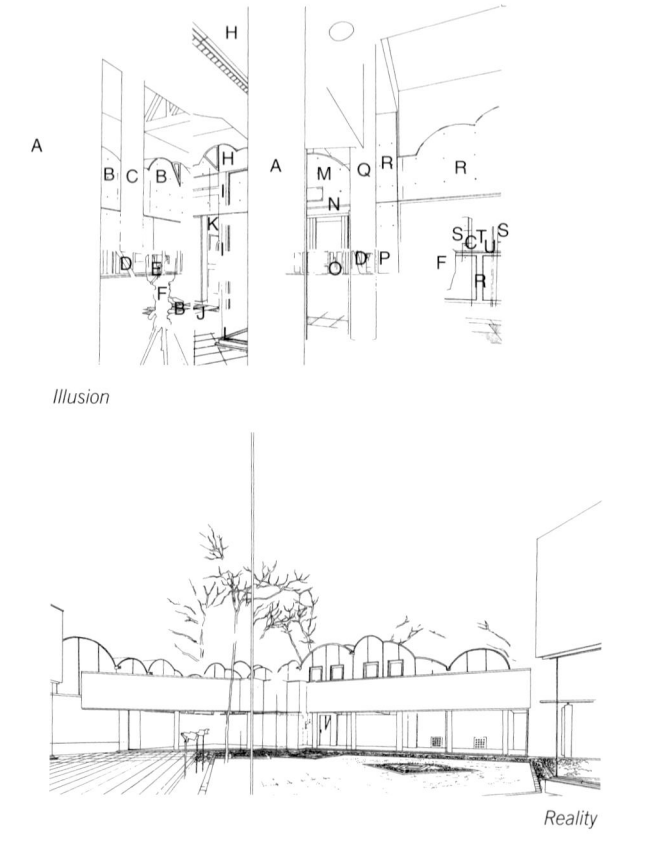

Illusion

Reality

consciousness. I have tried to explain the visual memory of consciousness using the phrase 'schema of scenery'. The schema of scenery is ambiguous, momentary and intermittent. Just as Marcel Duchamp explored consciousness in his artworks, I wished to investigate the function of consciousness in architecture. Architecture should retreat from Classical aesthetics; it should not only address efficacy but also human existence.

A combination of small and large-scale devices can present a completely new form of architecture. This new architecture is my ideal. To realise such revolutionary architecture, I paid attention to two rules. Firstly, that I must design with full attention to every corner of the building. Otherwise it would become dull and rough, with the device as its sole feature. Secondly, the building should be easy for everybody to understand. To achieve this, I used formal and utilitarian symbols that are common to all. Targeting the details of the newly introduced device, even at children, prevents complacency in the designer.

In the end, it seems that the meaning of 'unreal' in this context is 'real but at the same time not real', a contradiction, as we have seen, belonging to the logic of 'both-and'. I therefore have in mind a virtual rival, a computer program I will call ARCHITECT, which one may ponder as a device for inducing unreal architecture.

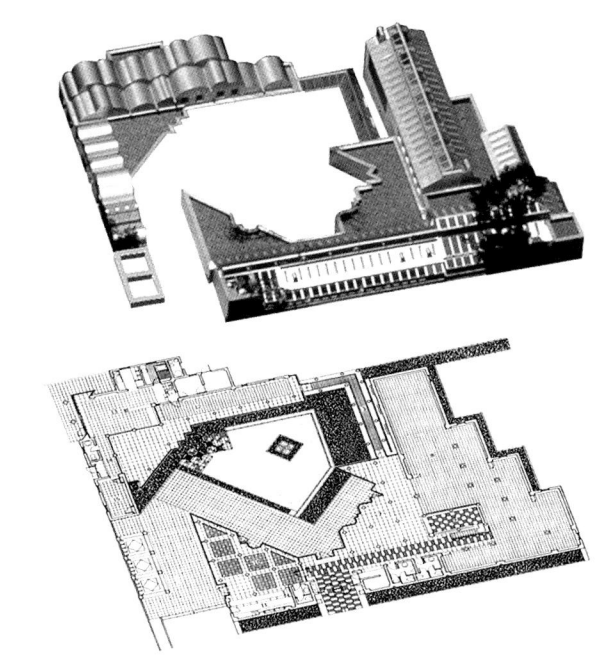

Tasaki Museum of Art: axonometric

Tasaki Museum of Art transposition photo: evening view of the courtyard from the lobby

Tasaki Museum of Art, Karuizawa, Nagano Prefecture, 1986

This small museum was designed to exhibit works by the artist Kosuke Tasaki. It is located in Karuizawa, a well-known summer resort, and closed in winter because of the cold weather. Natural light can therefore be used, especially as this is a private museum, unhindered by the constraints usually placed on a public facility.

The cloud-shaped roof was formed out of plywood, which was then covered with thin sheets of galbarium steel, a common roofing material made of steel, coated with an alloy mainly of aluminium and zinc. The cloud-shaped roofs in the later Yamato International and the Umeda Sky Building are made of aluminium panels. In the case of this museum, I depended on the skill of craftsmen rather than the power of machines to construct the roof. The details were also the work of the master craftsman in charge of the roofing. Electric heaters were installed in the valleys of the roof to prevent freezing.

It was in creating this building that I began to take the temporal dimension into account in design. That is, the concept of modality became clear to me. In designing a number of small buildings I had adopted the idea of 'mixing light'. I found that if I mixed light by introducing natural radiance from various angles and through variously shaped windows at the same time, the air in that particular spot became different in condition from the surrounding air. 'Mixing light' and 'designing air' have virtually the same meaning. I use the term 'modality' to indicate the overall condition of an interior space that changes in character from moment to moment as a result of the 'designing' of air, temperature and sound. The modality of architecture at a certain moment in time can only be understood as a changing condition. The objective of architecture is to capture clearly and to express accurately each of those moments.

A concrete element with a cloud-shaped roof is arranged in the lobby with a zigzagging glass wall. This is an instance of embedding and self-reference. A person traversing the lobby along this wall encounters diverse fictional scenes. *HH*

The zigzag surfaces of the glass, superimposing illusion and reality

Opposite: Bird's-eye view from the east

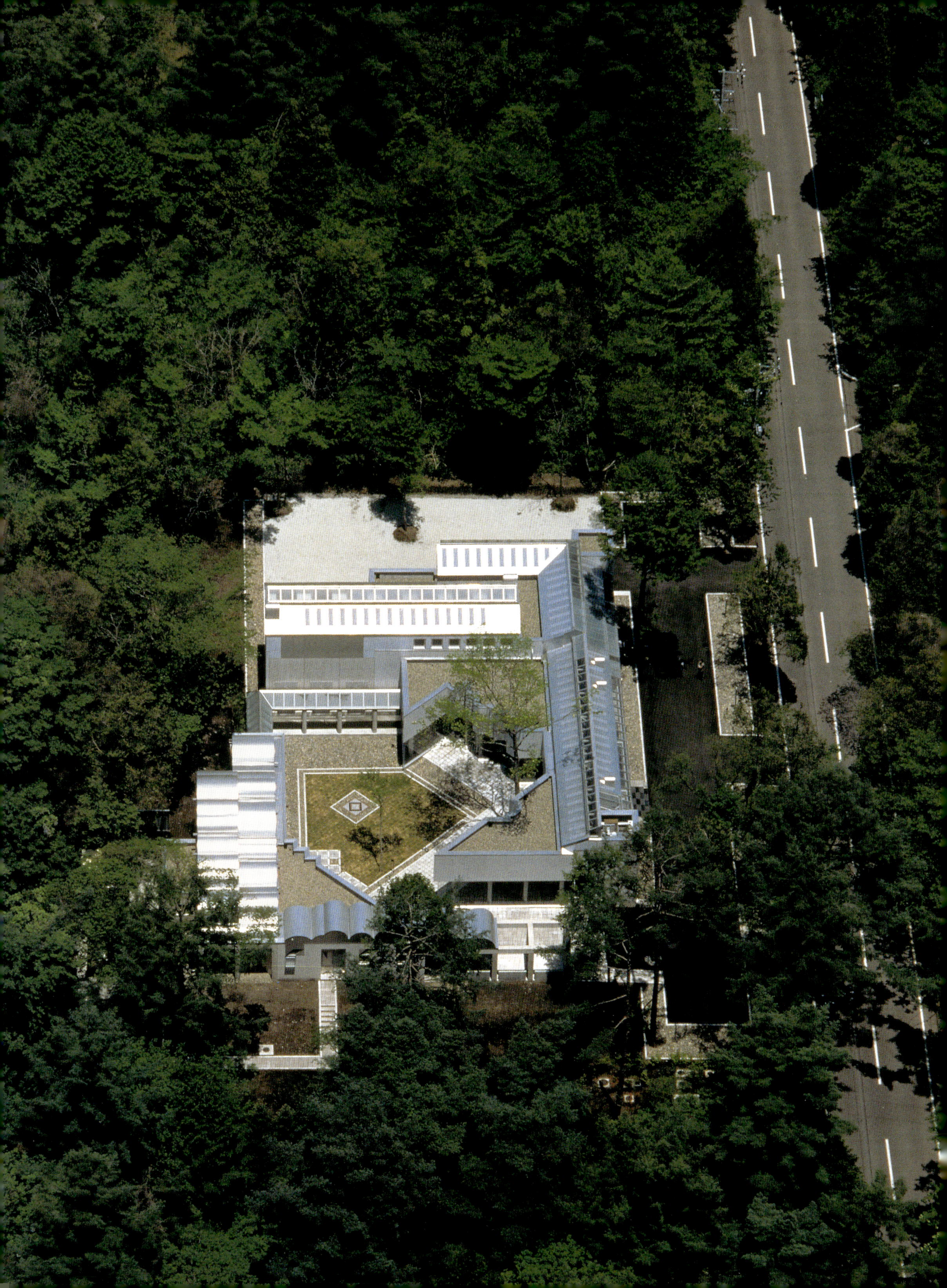

Hiroshi Hara

Detail of the south facade

Tasaki Museum of Art

Roofscape from the south

Overleaf: View from the south

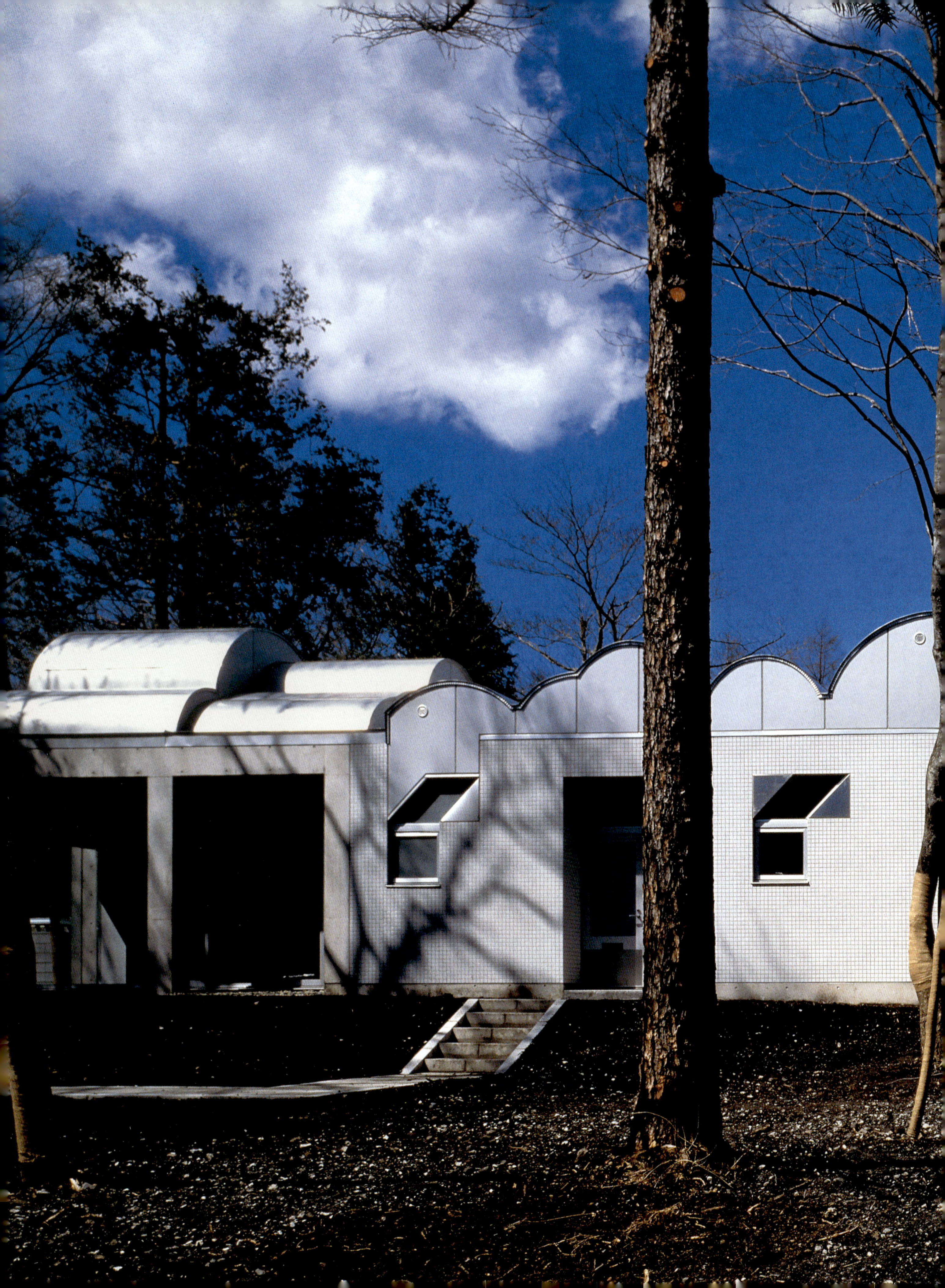

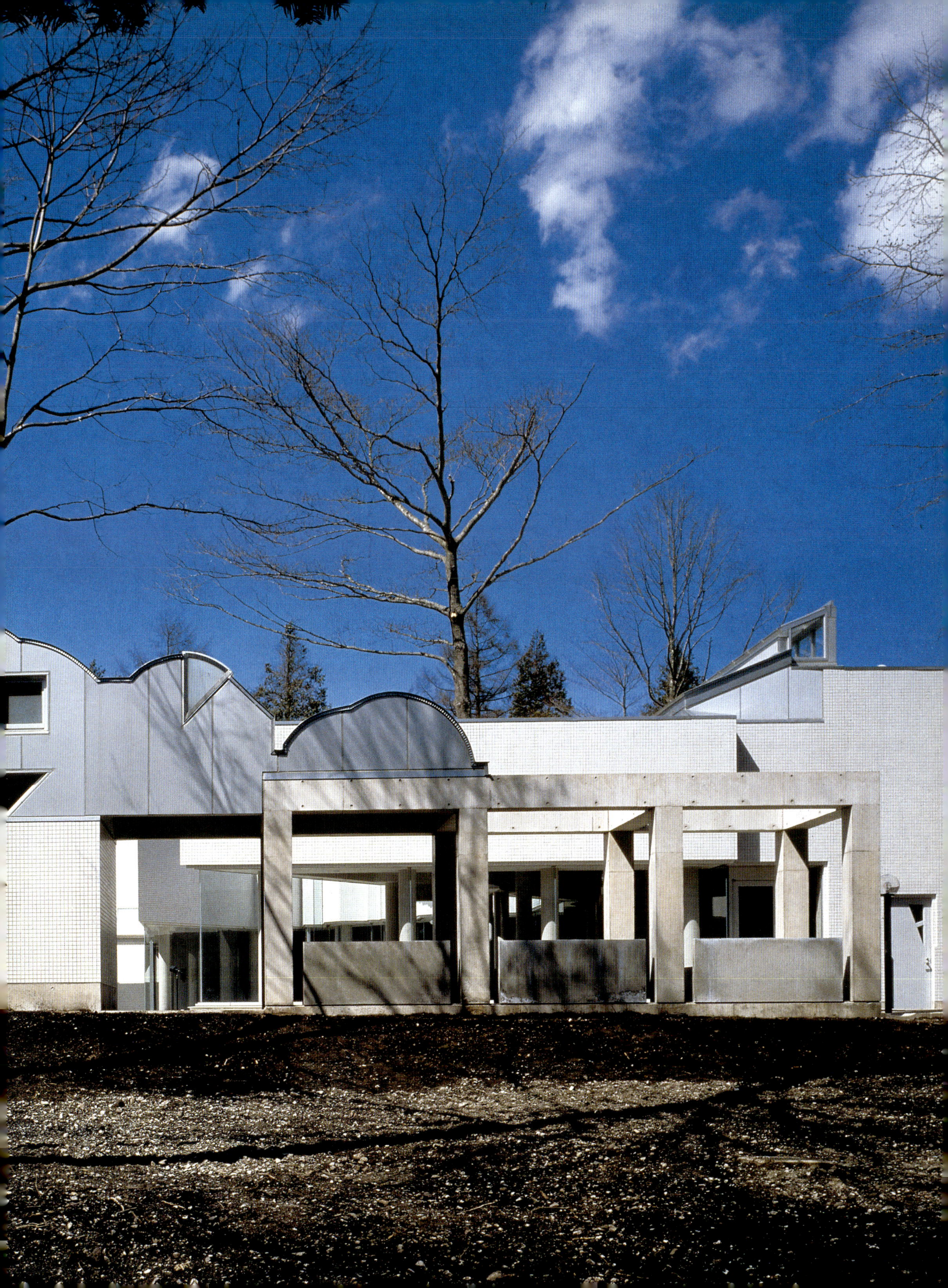

Hiroshi Hara

Detail of the south facade

Tasaki Museum of Art

Overleaf: Bird's-view from the southeast

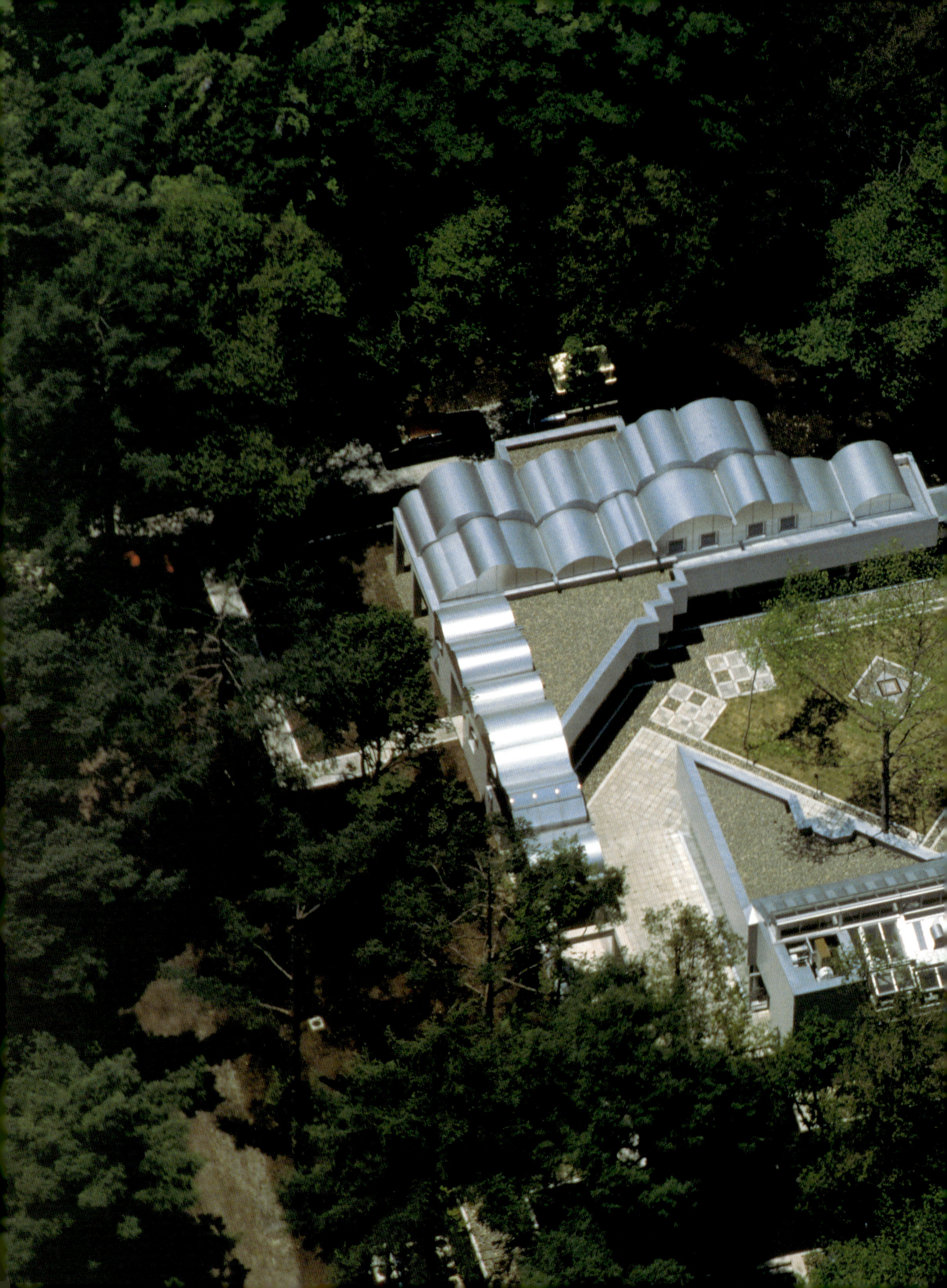

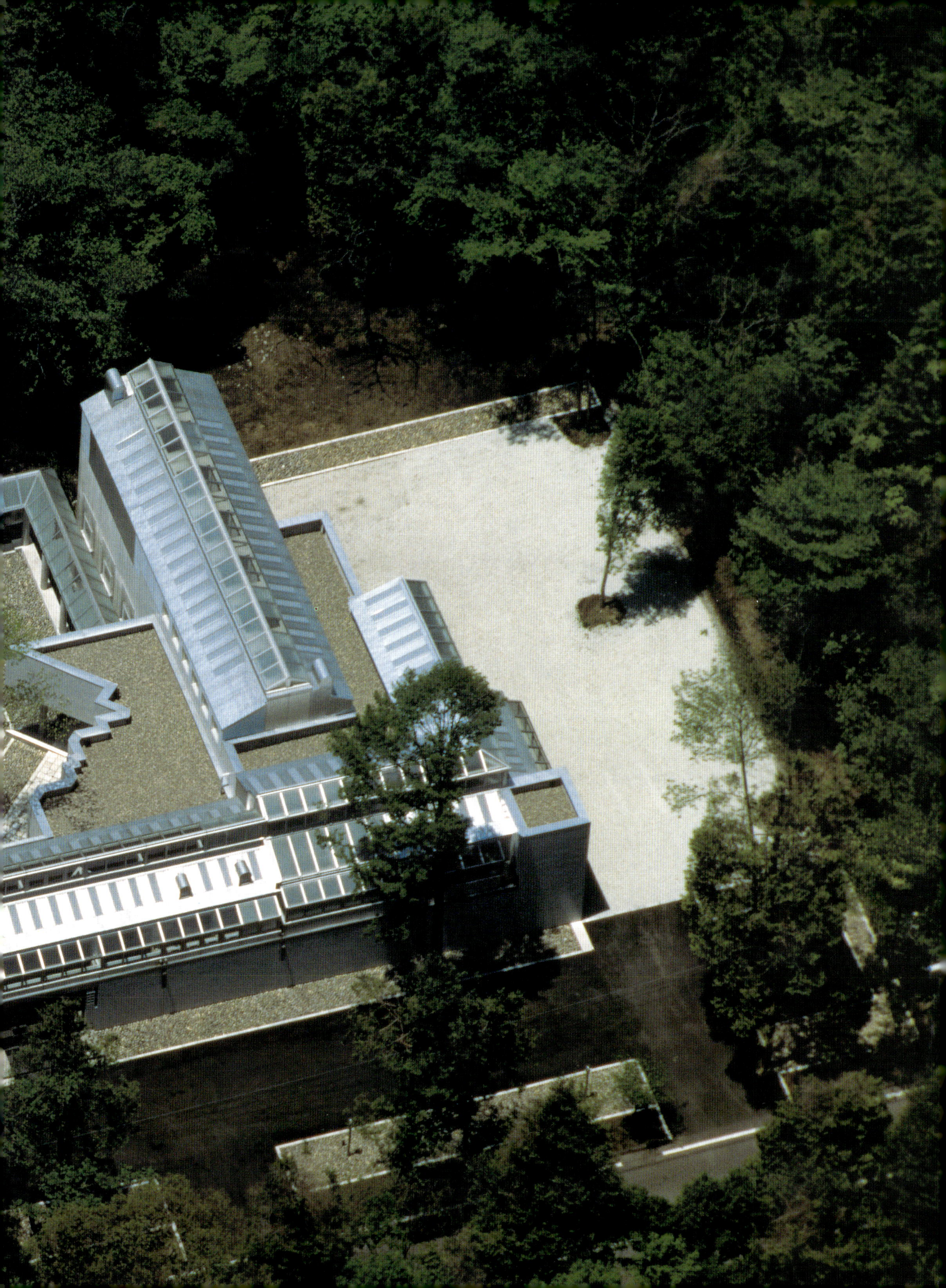

Hiroshi Hara

Tasaki Museum of Art

Upward view of the main gallery

Hiroshi Hara

Kenju Park 'Forest House', Nakaniida, Miyagi Prefecture, 1987

This house stands in the middle of the woods in a northern district of Japan. With its zigzagging glass wall, it is an extension of the design of the Tasaki Museum. Because it is surrounded by woods, I made 'transporting' the woods and sky my objective. I wanted illusionistically to recompose the woods and make a collage of the sky.

Kenji Miyazawa, a well-known writer of beautiful children's fantasies, was born and produced his creative work in this district. I have great respect for him, for he was a writer who showed us the sources of the human imagination. There are villages all over the world that are beyond anything most of us can imagine, and the fictionality of such villages is not unlike that of Miyazawa's stories. Like those villages, they exude human wisdom.

Miyazawa revealed how the woods have the power to stir the imagination. I wanted to express this power architecturally, although I had no definite narrative in mind. Why is such fictionality necessary in architecture? No doubt it has to do with the concept of place. If one of the tasks of architecture is to make a place distinctive (and I believe that it is), then various methods can be used for that purpose. However, when a place is reconstructed through architecture rather than left in its natural state, it inevitably takes on a kind of fictionality, whatever method is used. In that case, each place requires a different narrative; if the same narrative is simply repeated everywhere, all places will become similar and homogeneity will prevail.

Today, the act of dwelling in the woods is itself not theatrical. The pleasant space under a tree is one of the archetypes of architecture. In winter, the glass-defined living area becomes almost as cold as the outdoors. Thus in that season the living room retreats into the forest, becoming an extension of that archetypal space under a tree. *HH*

Views of the living area

Forest House

Hiroshi Hara

Forest House

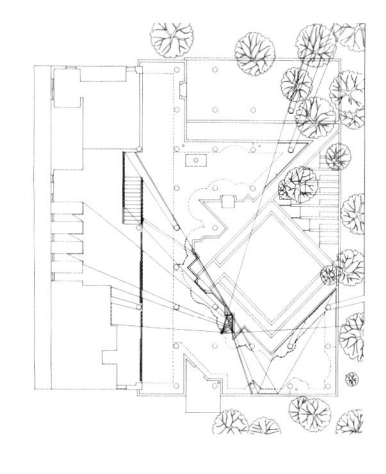

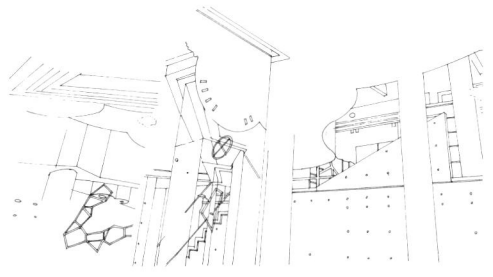

Above: Plan
Centre: Illusion
Below: Reality

Transposition photo of yard from inside

Hiroshi Hara

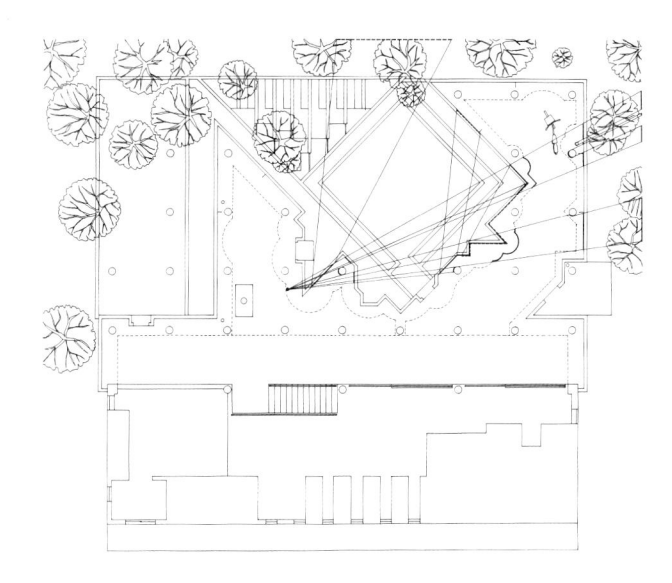

Forest House

Left: Plan
Centre: Illusion
Right: Reality
Opposite: Transposition photo of yard from inside

View of the glazed living area

107

Iida City Museum, Iida, Nagano Prefecture, 1988

I lived in Iida until I entered high school. The city is on a terrace of land in a valley lying between two mountain ranges. The beautiful natural environment of this region is the source of many of my ideas about architecture, including 'architecture as extension of the landscape', 'architecture as valley', 'reflection', 'designing air', 'corresponsive architecture' and 'modality'. Not only is the landscape beautiful, but the air is clean and the weather mild. As a child, I used to climb the mountains every day to gather firewood and edible wild plants, and to fish in the river. Later, in my visits to villages throughout the world, the inspiring scenes of buildings set in nature that I encountered made me realise the meaning of the landscape and of the lives people led in the valley in Iida.

Iida City Museum stands in front of a cliff on what was once the site of a small castle, where I often played as a boy. The lobby has a free form, while the gallery is uniform. It is a nearly closed space because it is intended to house, among other things, the works of Shunso Hishida, a Japanese-style painter who was born in Iida. Japanese-style paintings should never be exposed to direct sunlight or to unconditioned air.

The roof over the lobby is intended to echo the shape of the Akaishi Mountain Range. The museum exhibits both artworks and scientific displays, and the roof serves as a kind of topographic model. Steel plates were used to form its basic multifaceted shape. The master craftsman who roofed Tasaki Museum covered it with ordinary galbarium steel sheets.

Because of the historic nature of the place, it was decided to make the entire site, including the roof, a park. The studies of both Kunio Yanagida, the founder of ethnology in Japan, and of Konosuke Hinatsu, a man of letters, were rebuilt on the site. Both men had ties to this district.

A zigzagging glass screen separates the lobby from the front garden. Its effect is the same as the screens in the Tasaki Museum and the Kenju Park 'Forest House'. The facade can be seen from a distance. As in Yamato International, I designed it so that it would take on a bright golden colour at sunset.

I believe that children should be able to understand at least certain aspects of public buildings. Whenever I design a public building, I take into consideration the way in which children will experience it. *HH*

Close-up views of the roof

Iida City Museum

North facade

Overleaf: Bird's-eye view from the northeast

Hiroshi Hara

Iida City Museum

South view

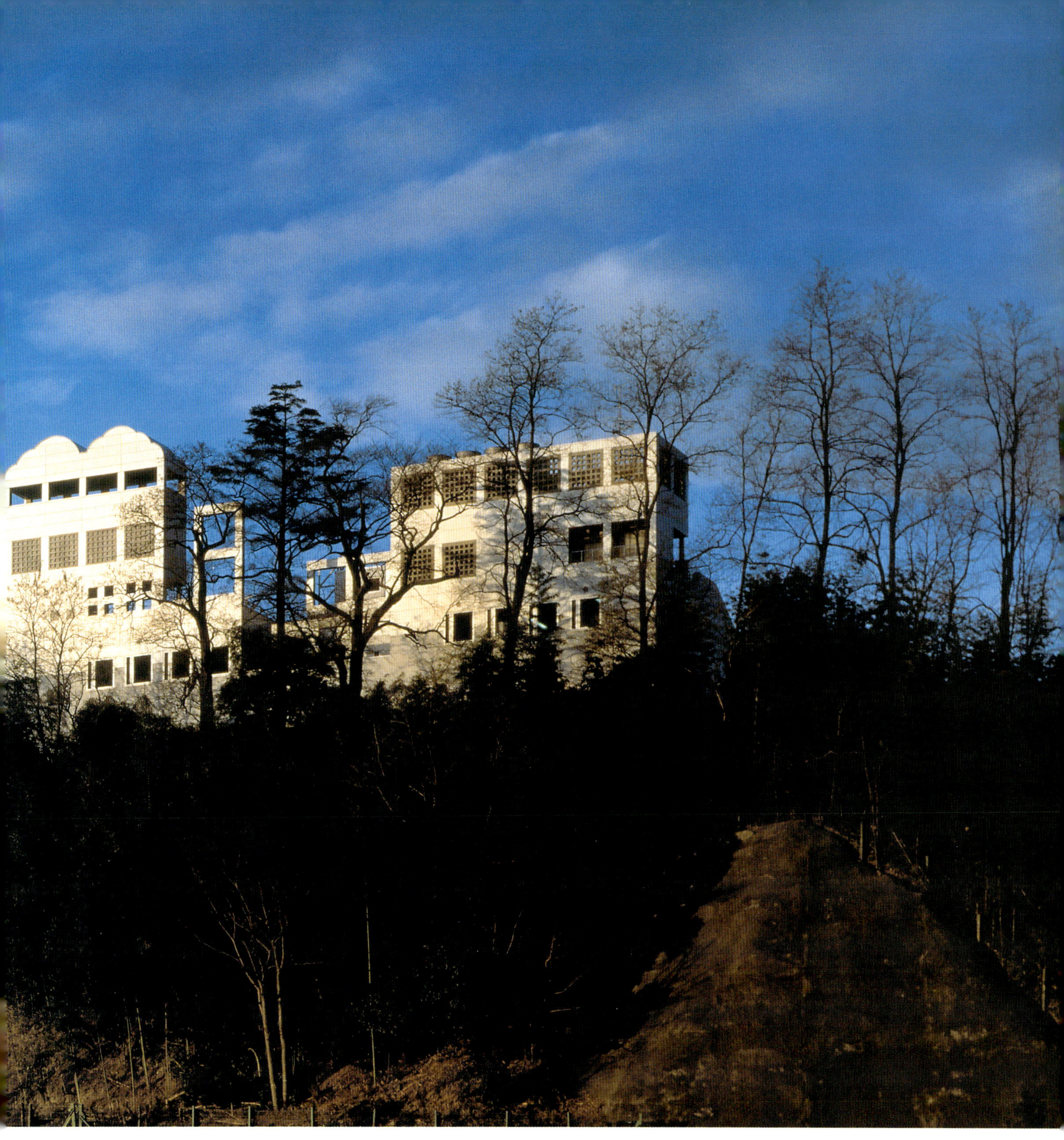

Main lobby

Opposite: View towards the entrance

Overleaf: General view from the southwest

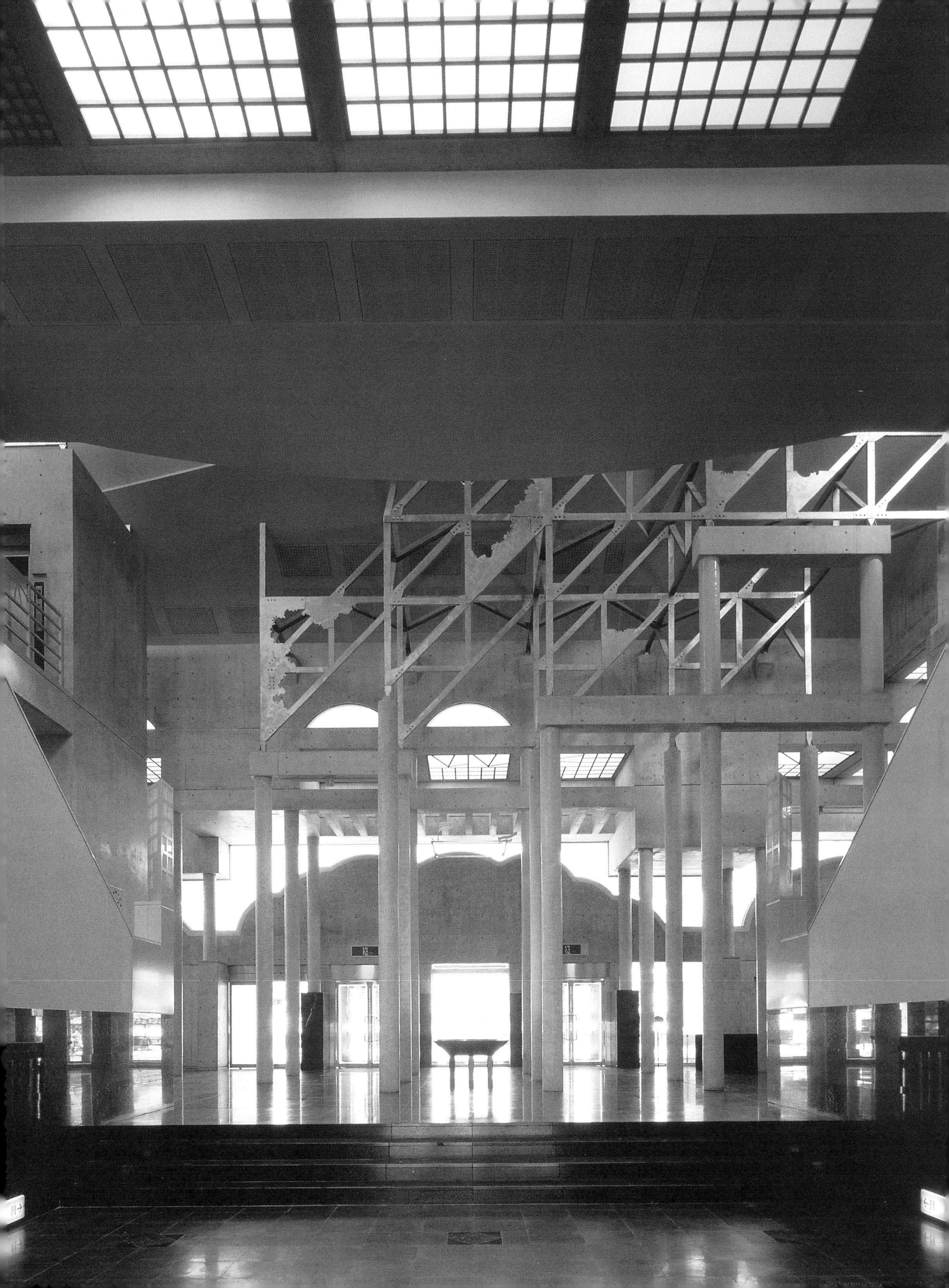

Hiroshi Hara

Iida City Museum

4. Floating / Mid-air City
by Hiroshi Hara

The idea of 'floating' occurred to me early on, when I designed my first work, the Ito House (1967). It has since become a major source for my architectural theories and techniques. The opposite of floating is 'fixed'. In terms of logic, the floating mode is a possible one, while the fixed is a necessary one. Floating implies the arbitrary or 'unplanned' light and hovering in physical aspect.

In the Ito House, the activities of the residents are somewhat pre-determined by the areas in which people and machines interact, such as the bathroom or kitchen. Other rooms including the bedrooms and the children's room also have such predetermined activities, since space is tight. These areas are considered as fixed. In the living room, however, residents can move freely in what can be considered a floating space. This area is covered by a wooden polyhedron roof. I made a Plexiglas model for this space in order to illustrate the movement of smoke, indicating that the floating area is a field of air. My intention was to let people 'move like the wind'.

I also made a model called the Induction House in which the rooms share an axle and each rotates in accordance with the sunlight. My intention was to show the concept of floating using this model. I now realise that only arbitrary rotation of the rooms would have achieved the true state of floating. That is, if all the rooms move together, they correspond to the concept of 'continuum'. If they move randomly, however, they correspond to the concept of 'individuum'. Floating therefore implies that things and people, as individuums, must move discretely. Conceptually then, in the floating area of the Ito House, each resident should move spontaneously.

Later, in the conceptual project for Montreal's 'International City' (1990) competition, I referred to street furniture as 'floating elements'. These are individuums placed along virtual parallel lines. They are urban tools that temporarily appear and disappear, different from the preserved historical buildings, which when partially modified are conceived of as chronological continuums. The street furniture, as individuums, and the historical buildings, as continuums, are contrasted in terms of time.

Floating implies clouds, winds and streams. Or perhaps wandering, or even freedom. As a movement, floating is a less constrained version of traversing; however, its literal significance, 'to float', still results in practical architecture.

The Umeda Sky Building in Osaka (1993) was based on the image of the hanging garden. Through my global field survey of villages undertaken in the 1970s as a research project at the University of Tokyo, I observed that the hanging-garden illusion, perhaps originating from Babylonian legend, is quite common in many places. Samara Tower near the Babylonian ruins, for example, is a spiral ramp for climbing to the sky, a good example of the hanging garden idea.

Similarly, in the Yucatan Peninsula, among the Mayan ruins, Tikal's tower protrudes above the jungle's treetops, affording an outstanding view.

Although the Umeda Sky Building is an office building, part of the requirement was that is should become Osaka's new landmark. I therefore proposed a skyscraper with a panoramic terrace. No ordinary skyscraper, it consisted of four towers supporting a connecting double-level platform. With the help of modern engineering techniques, I wished to realise a hanging garden that dominated the rest of Osaka as subordinate gardens.

In general, the Umeda Sky Building envisages the mid-air city of the future. Today's city, composed of highrise buildings, is expanding in 3D. Rather than concentrating on the vertical aspect, which creates a spatial dead end, the 3-D city depends on sky passages that connect buildings together at certain levels. In the early twentieth century, each building had its own underground level; now, these are mutually connected and became part of the wider city. The same thing could happen in the air. If public corridors were to be installed among the buildings, freer transportation would emerge.

The concept of the mid-air city is not original to our age; it existed in the early days of the twentieth century. The cul-de-sac reached by today's urban form, however, is the result of land and building ownership as well as administration. In order to actualise the mid-air city, some institutional modification is needed.

However, limited to small areas, the mid-air city should be feasible. An example can be found in the urban complex of the New Umeda City. After several changes, the hotel was eventually located in an independent building, connected only indirectly to the two forty-storey office towers. A 54 x 54-metre connecting part at the top of these towers, a bridge in the middle, a flying escalator and various terraces and elevator towers compose the mid-air city. Technically, this was difficult to realise in terms of rendering it typhoon- and earthquake-proof and I owe much to the construction company, Takenaka Corporation and others, and to Toshihiko Kimura, who helped me with the structural engineering.

The first phase of the construction was to build the twin towers. Next, the connecting part was assembled on the ground and lifted up to the top, accompanied by a ceremony and TV broadcast, shown throughout Japan. It provided a good opportunity to inform people about modern construction technology.

The technology and design of the mid-air city are also applied in Kyoto Station (1997). Though only 60 metres high, it is a 470 metre-long urban complex. As in Osaka, bridges and terraces create the atmosphere of the mid-air city. In this building, people can choose certain paths and wander in a variety of directions. They are therefore

Floating / Mid-air City

'Extra-Terrestrial' Architecture: Lagrange Point and the Image of Space Potential

'floating' in two ways: first in terms of 'moving in mid-air' and second in terms of freely selecting their own paths.

Having moved on to projects such as skyscrapers and urban complexes, I began to consider further the possibilities of the concept of floating, the mid-air city and the large-scale building. Envisaging a 500-metre cube, I asked, if this were considered as an urban frame, what kind of possibilities would arise? I made many drawings and models and discussed with my colleagues at the university the impact that this 500-metre cube could have on the environment, in contrast to the dispersed residential forms in Tokyo. The results, however, were inconclusive. One thing that can be definitively stated is that even in this gigantic cube, people could move between two locations within ten minutes, which might alter the character of current urban facilities, including administrative elements. Urban areas, not only in Japan but also in many other Asian countries, continue to spread. Since this is bad for the environment, especially in terms of the decrease of farm land, I believe the proper solution is to reform the existing urban area, not to expand it. In the latter half of the twentieth century, architects lost the power to conceive new urban solutions. From now on, I believe we should propose new images of the city from the environmental point of view.

After this exercise in enlargement, by mere coincidence I went on to think about architecture outside the Earth. Working with space-science experts, I designed structures to be built in outer space, mainly on the Moon. I named such structures Extra-Terrestrial Architecture, the ultimate in 'floating' architecture, since gravity is low.

All things considered, to live in space under current technology is difficult though not impossible. This is even true of the Moon's surface. People would have to live in heavily equipped underground houses. The temperature would differ by about 300 degrees Celsius between the sunny surface and the shade. Lack of air and water as well as various problems of radiation add to the difficulty, making it unrealistic for humans to stay long on the Moon. Without the discovery of cheaper rocket fuel, or new methods of transportation, permanent residence on the Moon would be unrealistic. Of course, space travel is another story.

From the moment we entered space, it was important to recognise the preciousness of the Earth. Reaching North America in the age of discovery and landing on the Moon were completely different experiences in terms of understanding this. What Extra-Terrestrial Architecture taught me was that the continuation of human history equals the preservation of the Earth's environment. In the future, it is certain that Extra-Terrestrial Architecture will be built, at least in the realm of the imagination, and this is an interesting scientific theme.

From the virtual world of the 500-metre cube and the Extra-Terrestrial Architecture, I returned to the real world on a smaller scale with the design of five modest residences (Cube Houses, 1998). These take the form of cubes sitting on a plane. In a mathematical

First sketch of Umeda Sky Building (1988)

Floating / Mid-air City

progression, they ranged from 5.2, to 5. 8, to 6.4m, to 7.0m and finally 7.7 metres in dimension. Only the largest one was not built. While working on the 500-metre cube, I had imagined that if this kind of city were to be realised, it must be designed by a virtual computer software program that I called ARCHITECT, which could handle the enormous number of possible answers. When I began work on the tiny cubes, I used my own design sense, but continued to ask myself, how would ARCHITECT design them? In this case, I conceived of ARCHITECT as a novice-level computer that had knowledge of modernist houses.

The basic design concept of the Cube Houses is discreteness. Architecture based on discreteness is not continuum, but individuum. While the old type of neighbourhood community is a continuum, today's technology potentially connects discrete elements to form a community, regardless of distance. Architecturally, what would a discrete city be like? The four houses, perhaps designed by ARCHITECT, were built on distant sites, though they are all situated in Japan. I intended them to 'float' discretely 'floating'.

Currently, I am working on three other houses. This time, the dimensions are determined by geometric progression and are applied to the gardens, ie, a discrete design of void attractors that potentially offers a new development in terms of floating and discreteness.

Sketch of Mid-air City (1989)

Ito House, Mitaka, Tokyo, 1967

The Ito House is one of my first buildings. It has been preserved in its original state in all important respects, despite a later addition to the bedrooms.

It represents my starting point in a number of ways. Firstly, I designed it with the concept of a 'floating' domain in mind. ('Floating' would become the source of many of my subsequent images.) I was attempting to create under a polyhedral roof a free domain that would place relatively few constraints on the action of the occupants, as opposed to those domains (such as the bedrooms, kitchen and bathroom) where the types and forms of action were easily determinable.

Secondly, I also had in mind the concept of 'fields' when I was designing this house. This fact is clear from my diagrams showing brightness and the movement of air. The design shows a clearer understanding of architecture as space than the later Reflection Houses. Although I was not aware of it at the time, the lighting device made of wire-mesh that I designed for the living room seems to serve, together with the central column, as an 'attractor'. If that is the case, the Ito House shows an awakening consciousness on my part of 'semiotic fields' and helped shape my present conception of architecture as space rather than as thing.

Thirdly, the house is a polyhedral form composed in a random manner. Later, I was to design polyhedral forms, each composed randomly out of triangles and quadrilaterals, on the facade of Kyoto Station. The important thing I discovered here is that any form can be created through a combination of planar elements, no matter how irregular it may be. This method of formation leads to a geometry with a will of its own, so to speak.

For these reasons, the Ito House is for me an important work. Computers were unavailable then, and I designed the roof of this house using measurements from a model. *HH*

The living area

Ito House

South facade

The World of Yukotai

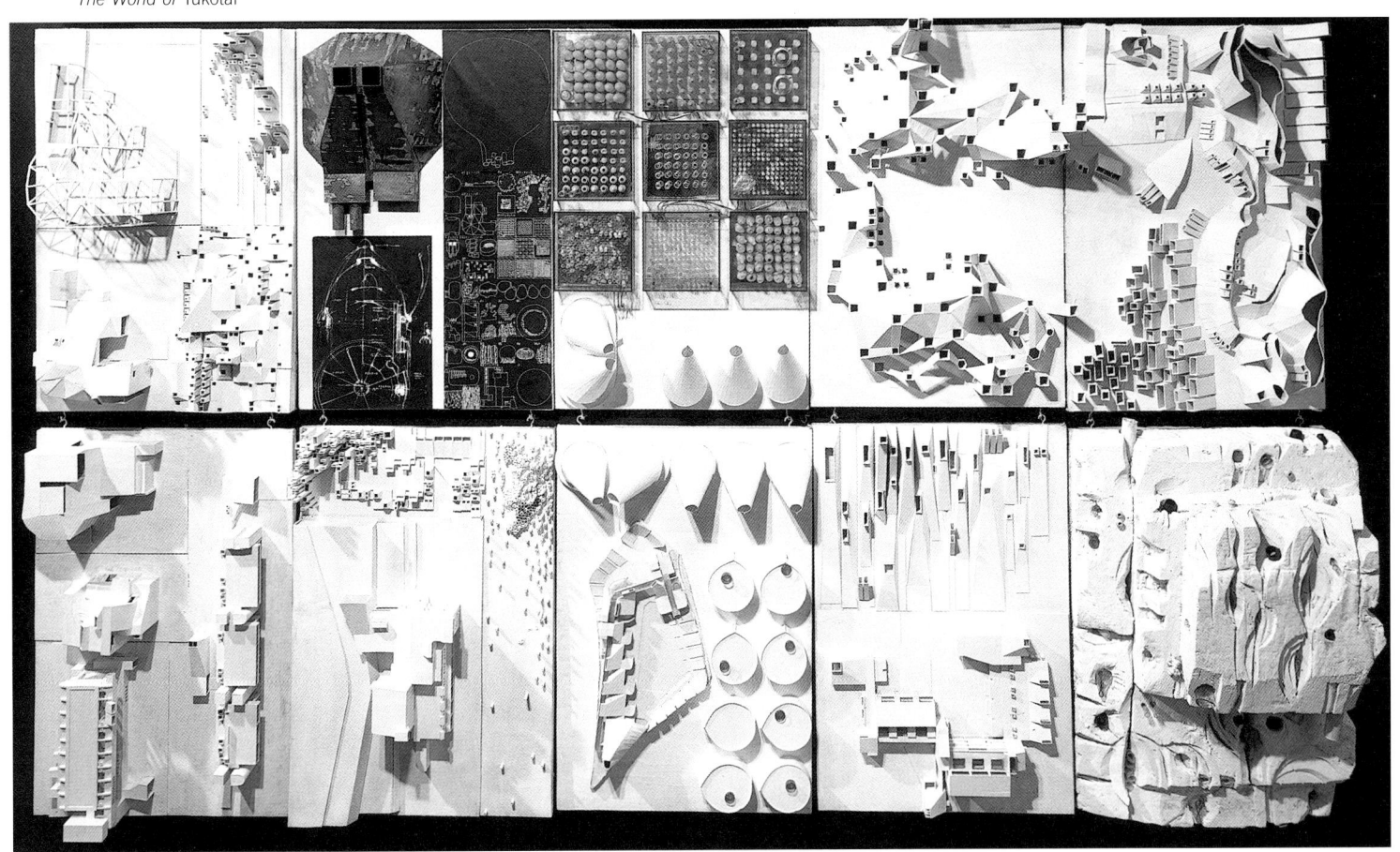

125

Hiroshi Hara

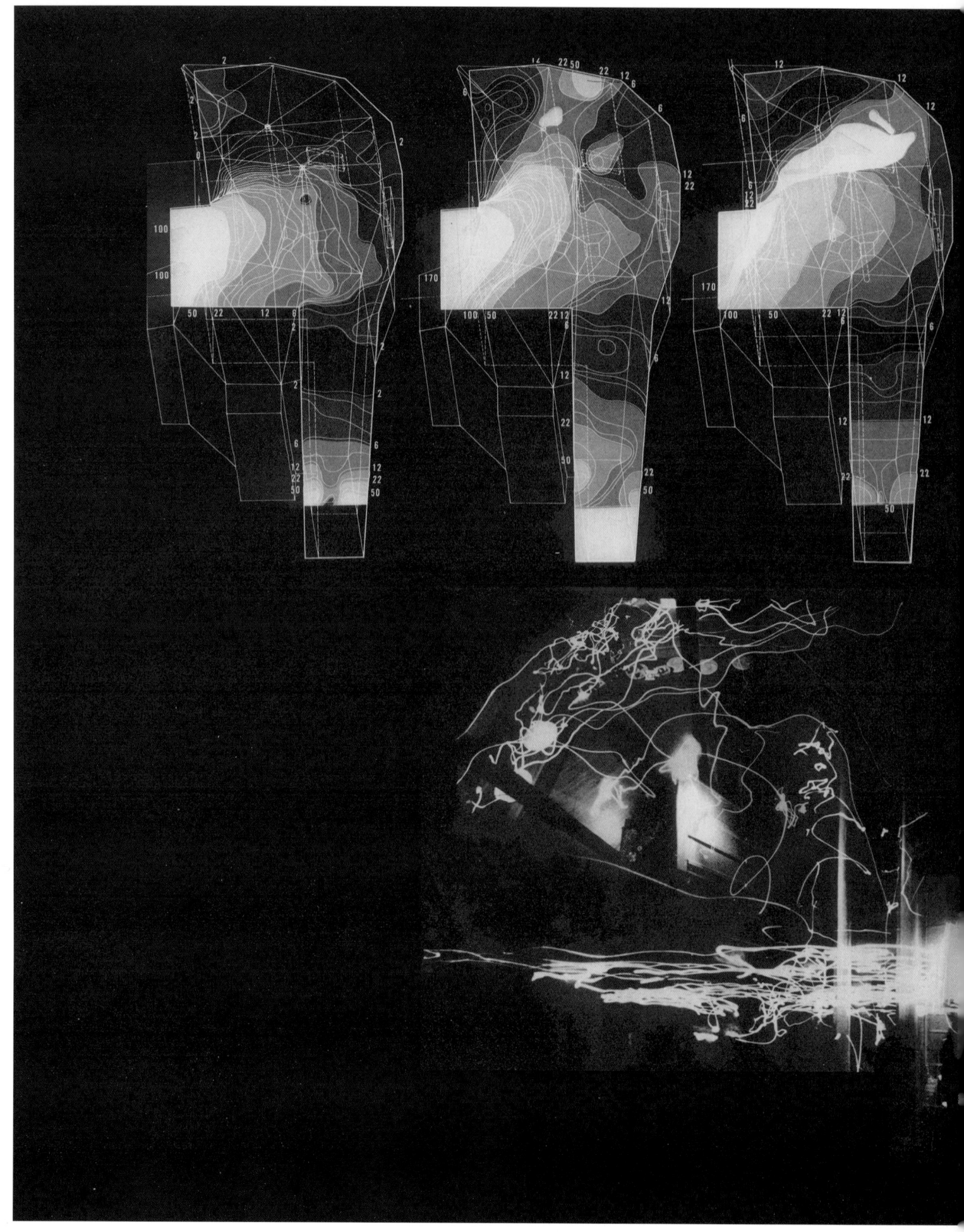

Ito House

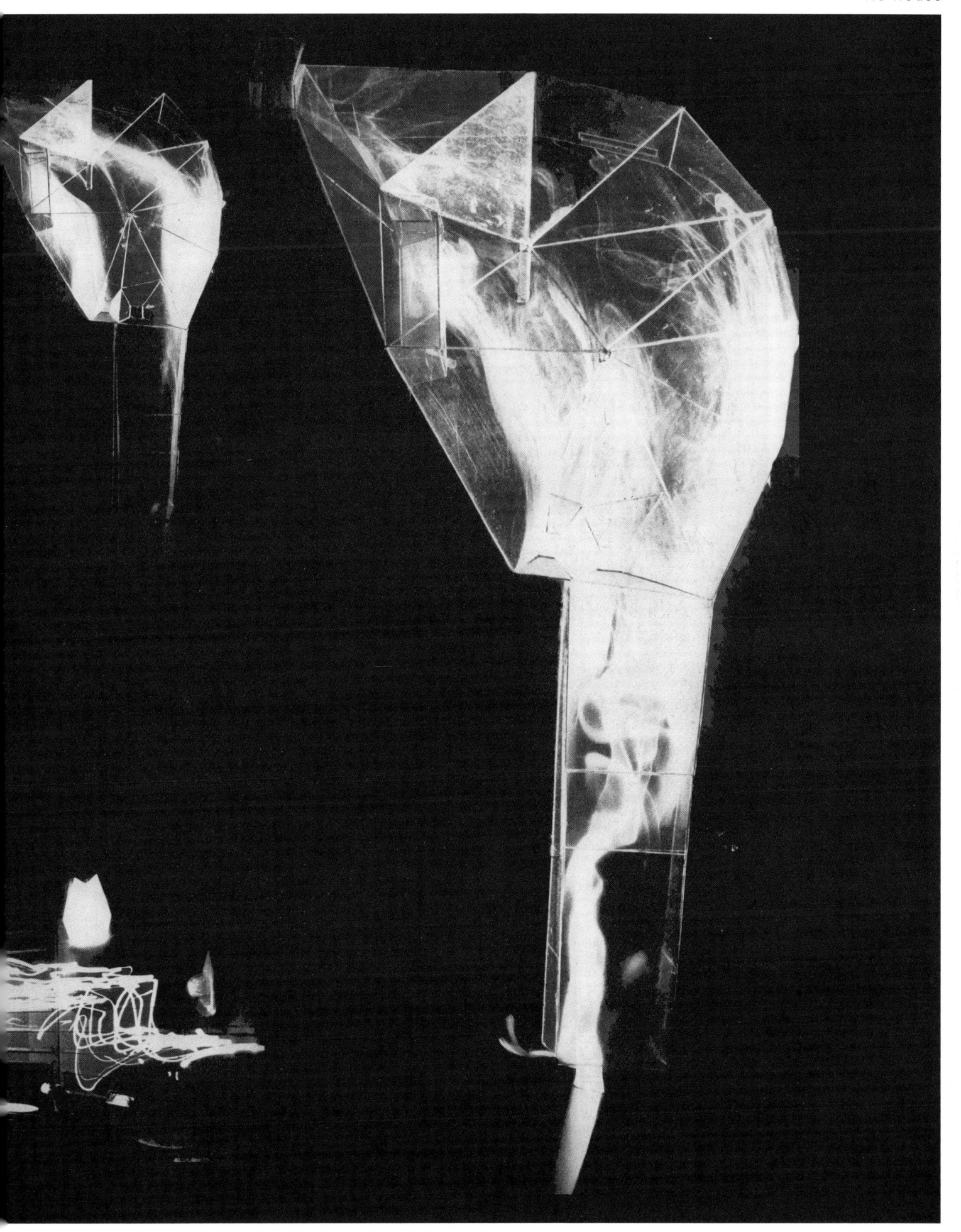

500 x 500 x 500-Metre Cube (Theoretical project), 1992

This 500-metre cubic framework, with a volume beyond the imagination, is a Pandora's box for examining the possibilities of a new urban structure. Essentially a three-dimensional, mid-air city, it is the result of extensive trials. The presentation below of one of its possibilities is a qualitative rather than quantitative study that combines the concepts of 'Burst and Repair' and the management of these concepts within a structural 'frame'.

Firstly, a 1:500-scale framework as a structural *tabula rasa* is constructed and used as a three-dimensional grid. This can be employed for plans, elevations and sections but, in the end, it was used mostly as a section. This allowed for 'Burst and Repair', in other words the possibility of the simultaneous existence of heterogeneous factors.

Although there are many ways to develop diversification, we layered and combined images after reviewing twentieth-century urban theory. During that century, there was first a simple yearning for nature, in contrast to the machine aesthetic that replaced it later, when concerns about industrialisation and community building gave rise to various manifestos and 'isms'. At the same time, the concept of homogeneous space was developed. Ironically, this universal idea is similar to those espoused by Fascist regimes in the middle of the century. After the 'burst' of the Second World War, the idea was used as a tool for the development of cities around the globe. Simultaneously, however, many subcultures were flowing into the city. These heterogeneous elements existing within the homogeneous space constitute the fear of Modernism. Postmodernism's clarion call, a break from Modernism, is a means to escape this fear, but it remains to be seen if its doctrines can be incorporated into such a megastructure.

The machine age has suddenly turned into the electronics age. Electronic technology used for operational practicality has infiltrated the city to a degree, but stops short of altering its architectural or structural form. Soon, the call for a 'third nature' will provide another 'burst'.

The homogeneous space in our project is a vessel for these tautologies, inside which float the accumulations of various co-ordinates. Regrettably, our study has not so far produced any breakthroughs in terms of a proposal for new urban structures. However, in the end, I did conclude that one of the principal attractions was the immediacy of urban facilities, wherever one was within the structure. As a next step, I plan to examine this aspect more closely. *HH*

Photo of model: 500Mx500Mx500M-1993

Opposite: Photo of model: 500Mx500Mx500M-1992

500 x 500 x 500-Metre Cube

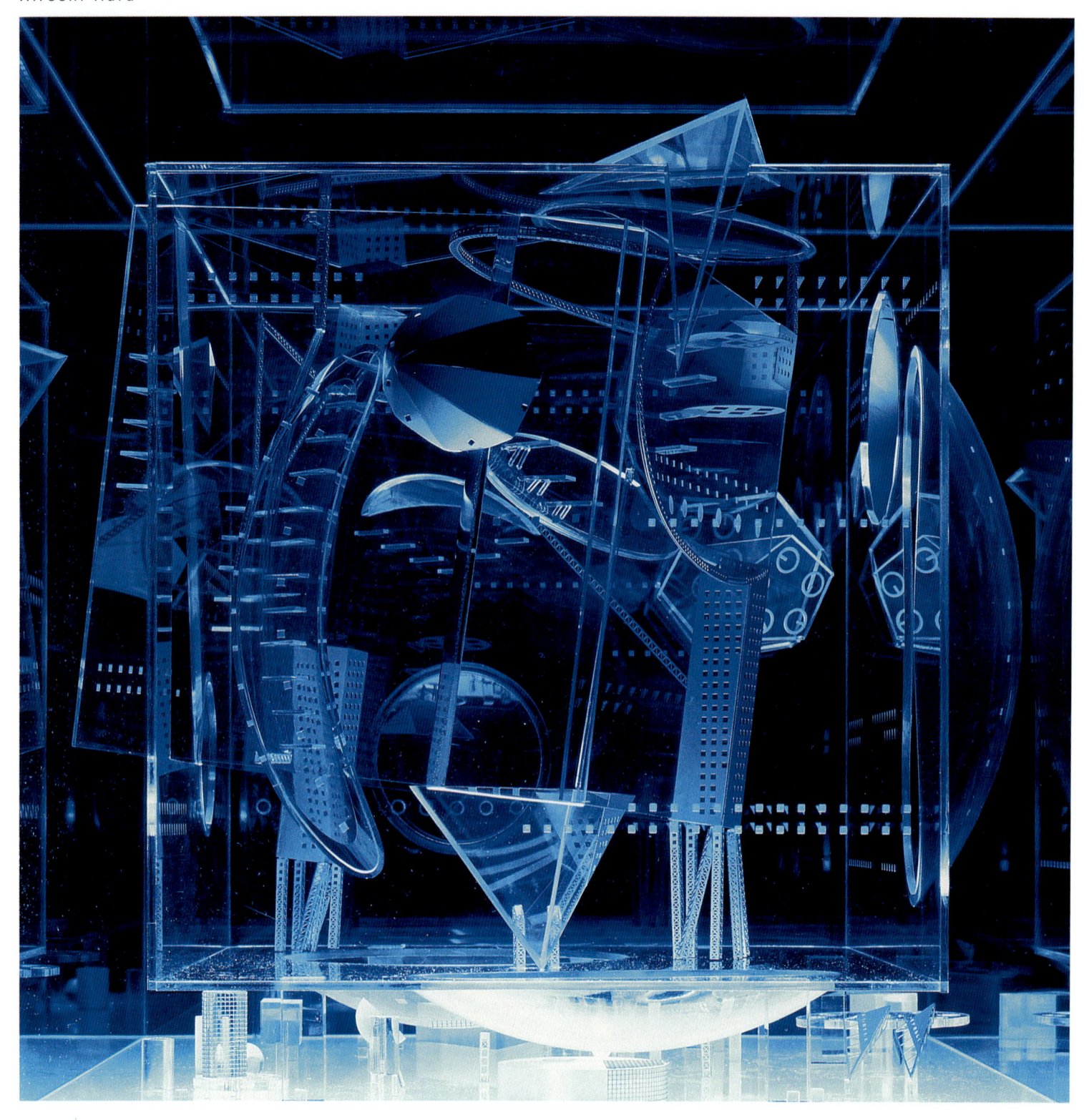

Hiroshi Hara

Model photo: 500Mx500Mx500M-1993

500 x 500 x 500-Metre Cube

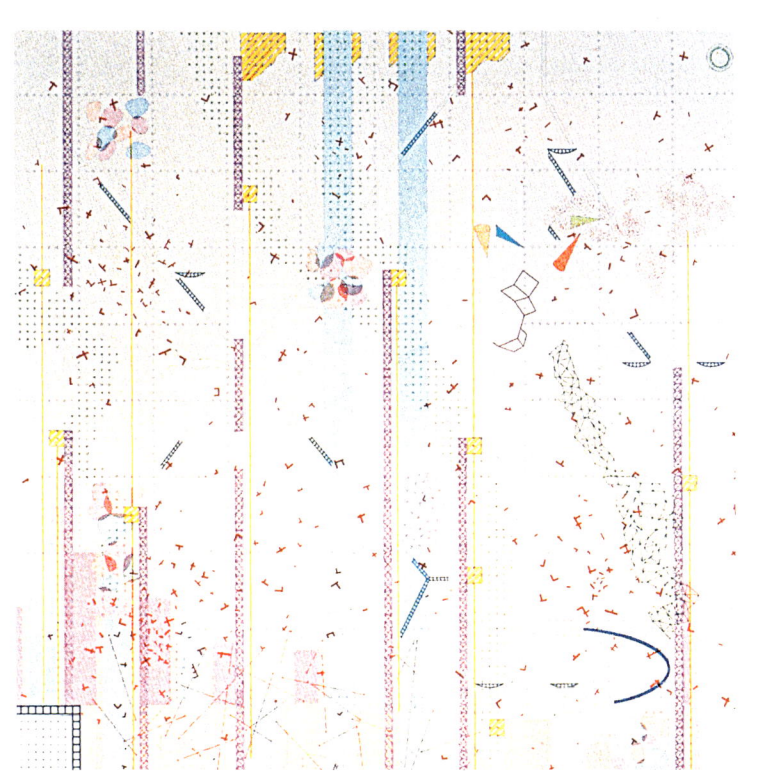

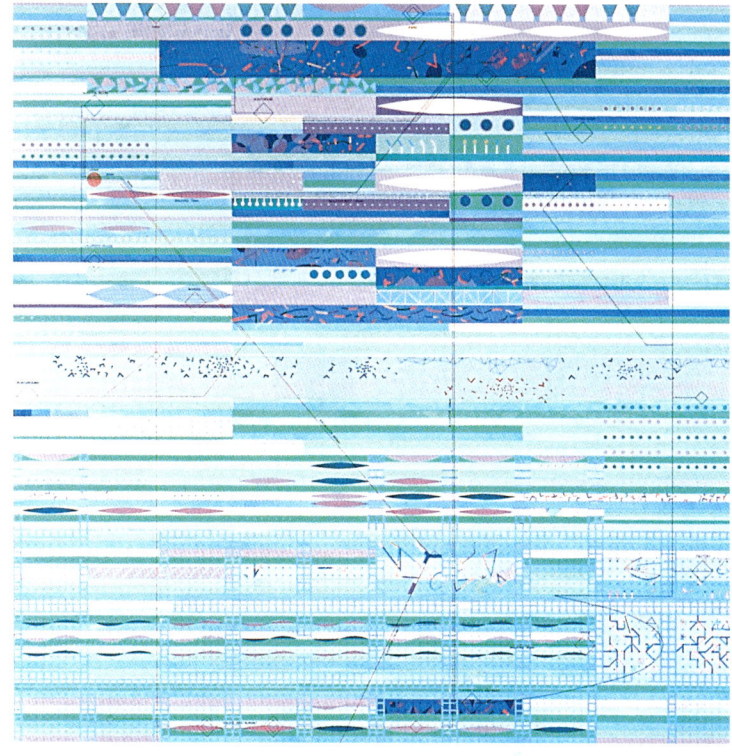

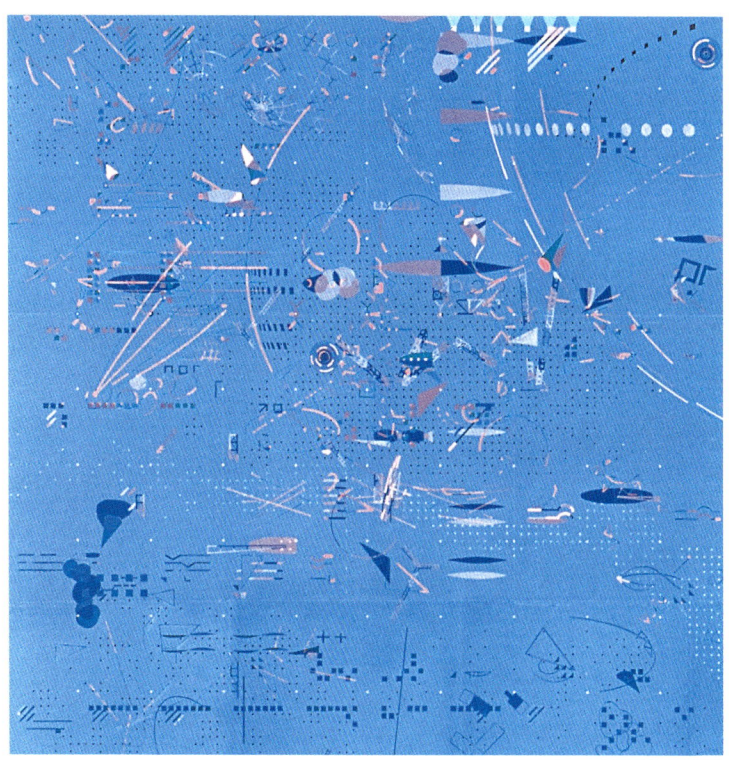

2041   *2081*

2091   *2101*

131

Umeda Sky Building, Osaka, 1993

This 'connected highrise' structure is comprised of a pair of office buildings whose design was chosen from a number of international proposals. The clients were Sekisui House, Toshiba, Daihatsu Corporation and Aoki Construction Company.

The site is in the central district of Osaka, but in an area that has not yet been fully developed. One of the design conditions was that the buildings should be able to generate a high level of human activity.

The history of tall buildings and surveys of villages show that people of many different regions and periods have envisioned hanging gardens. Here, I proposed three connected highrise buildings with an open-air observation deck on top. In the end, the hotel became a separate structure, leaving two connected highrise buildings.

My first sketch showed four buildings supporting a platform with an aperture. Connecting highrise buildings with a large deck, not to mention bridges, is difficult in Japan because of the threat to the structure posed by frequent earthquakes. The realisation of these intentions was made possible in large measure by the work of Toshihiko Kimura, the structural designer. I felt that something more was necessary to create the effect of a city in the air. The highrises would need to be at least forty storeys high and have an open-air space in the form of a plaza. The plaza floating above the central district would then become a hanging garden from which to view the city below. Hanging gardens in history have not literally been gardens; they have been envisioned instead as devices for scenically recasting the spaces around them.

Central districts made up of highrise buildings are not truly multidimensional in their present form. This is because, from an urban planning perspective, each highrise building forms a cul-de-sac. In the Umeda Sky Building, the intention was to develop and install a set of equipment – including a plaza and mid-air escalators – that will make possible the realisation of a truly three-dimensional city.

The deck at the top consists of two storeys of spaces and the rooftop. This element was assembled on the ground and raised up during construction. *HH*

North-south section (basic design 1989)

Opposite: Upward view from the south. The 150-metre-high atrium space is crowned by the sky garden. The sky elevator rises vertically

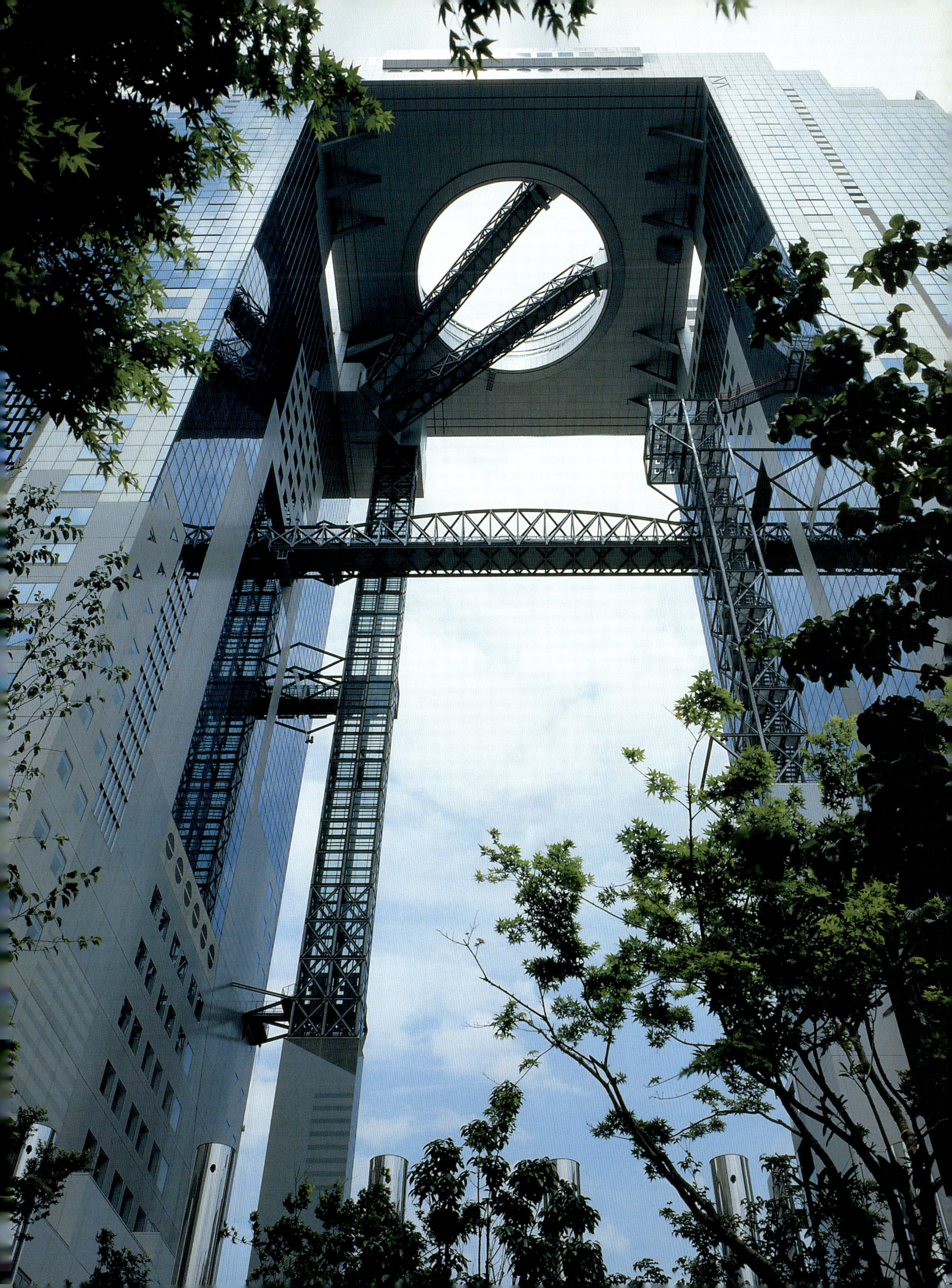

Hiroshi Hara

General view from the southeast

General view from the north *Overleaf: Upward view from the foot of the annex building*

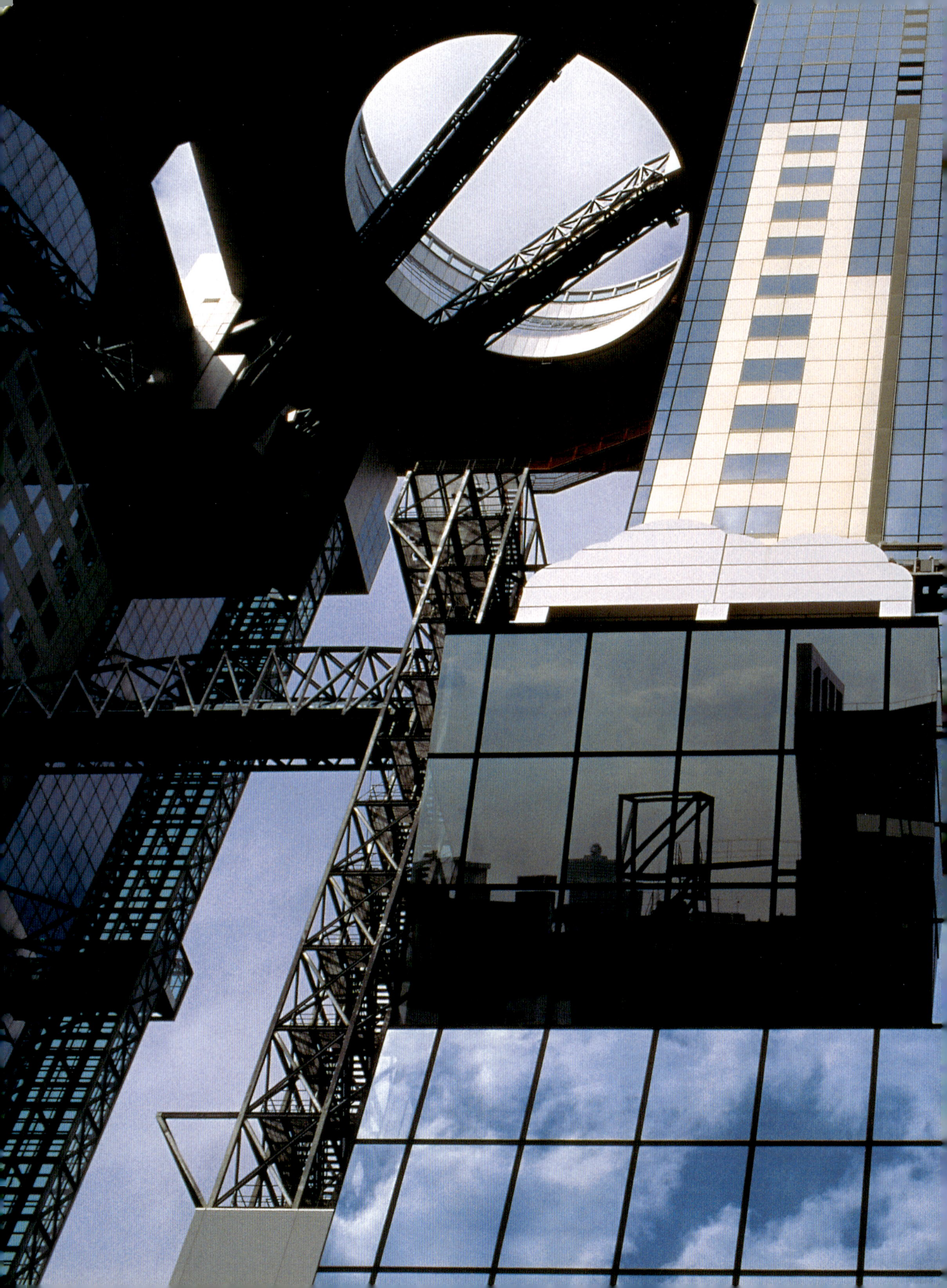

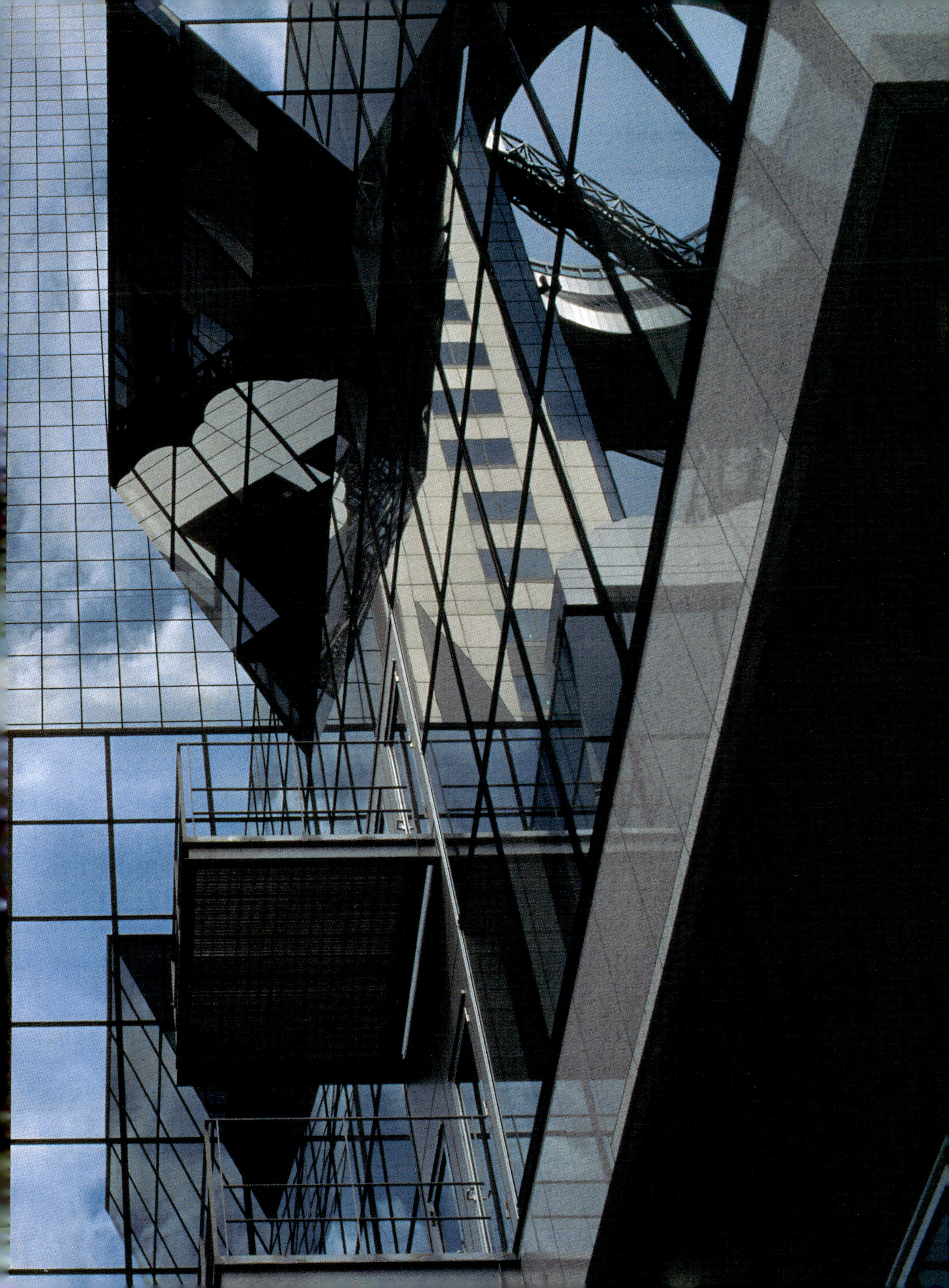

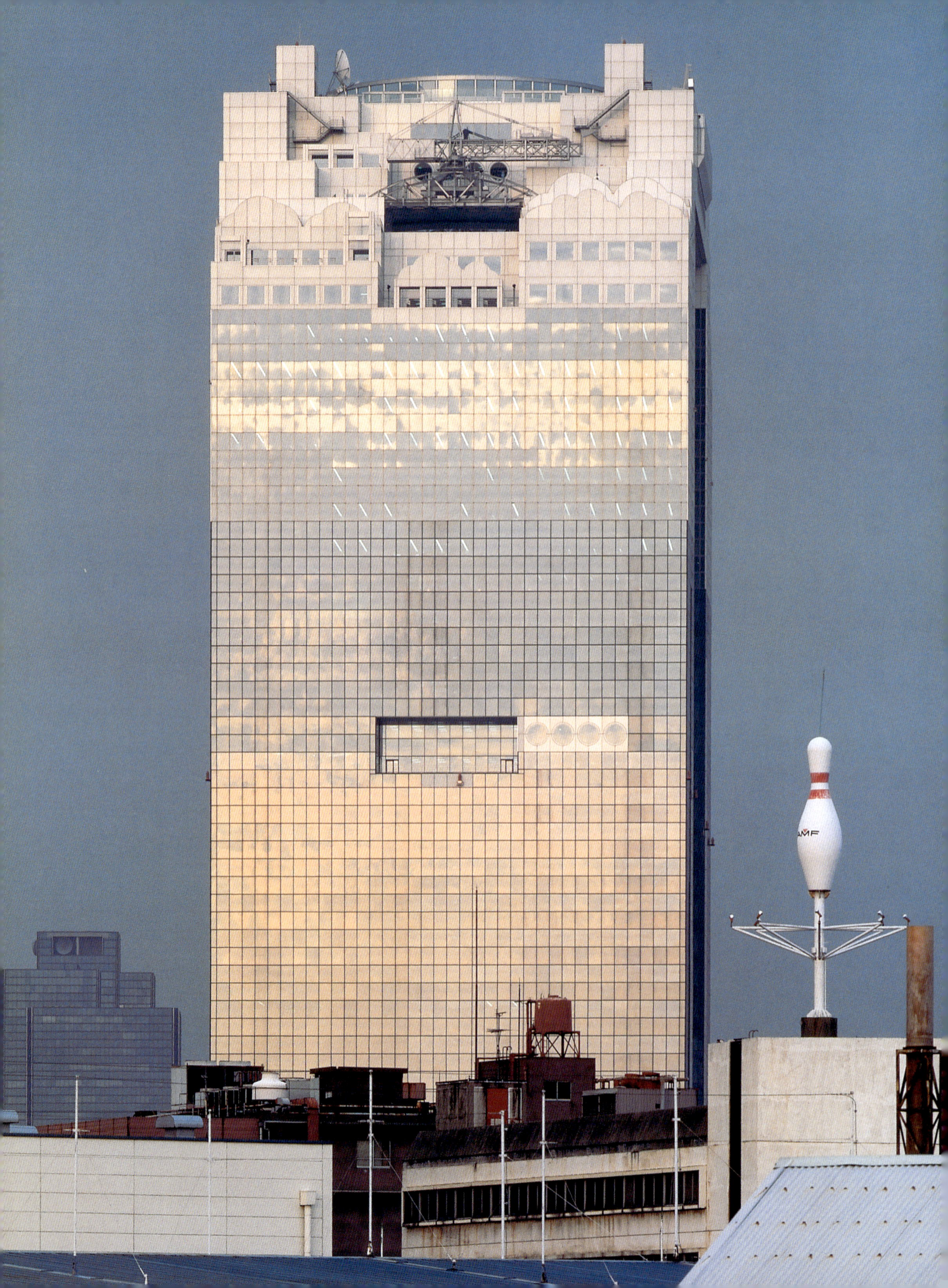
AMF

North exterior in the evening

Opposite: General view from the west

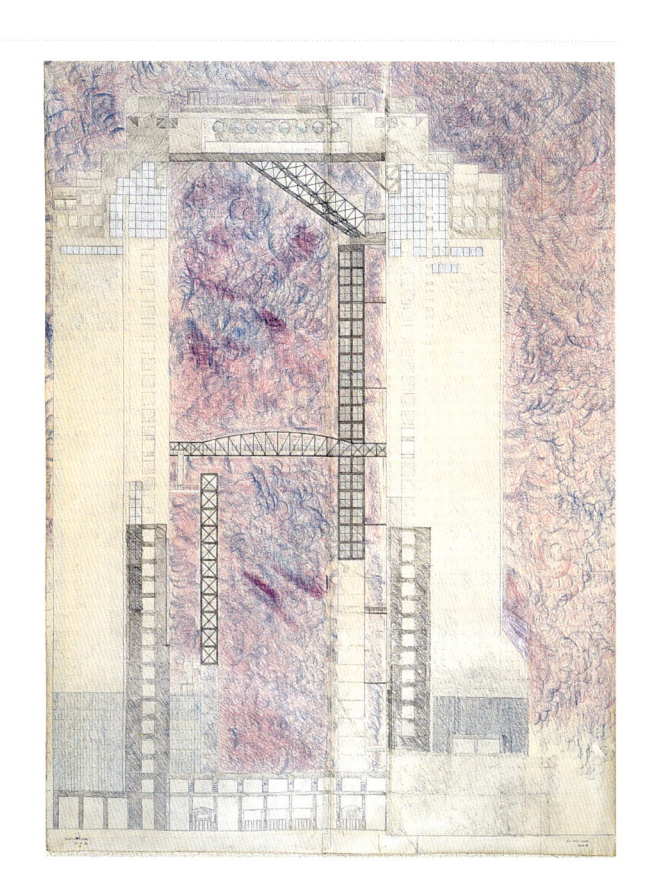

North elevation (basic design, 1989)

Looking up towards the sky garden

Opposite: Schematic facade sketch of three wings joined together at the top (1988)

Overleaf: The sky deck as hanging garden

Close-up view of Mid-air Garden

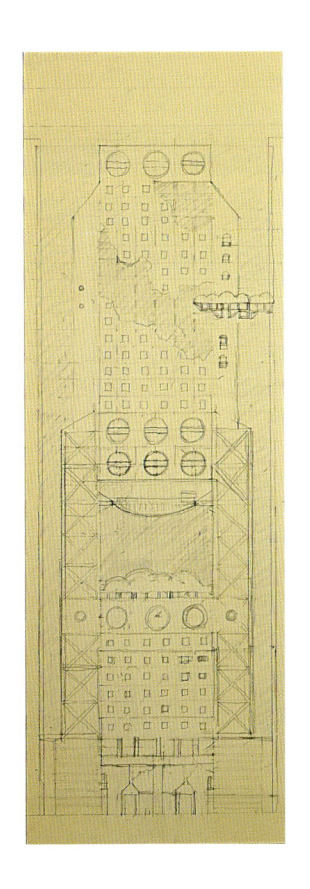

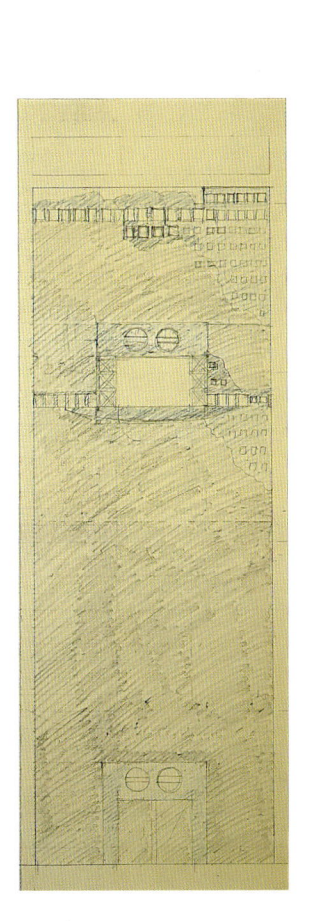

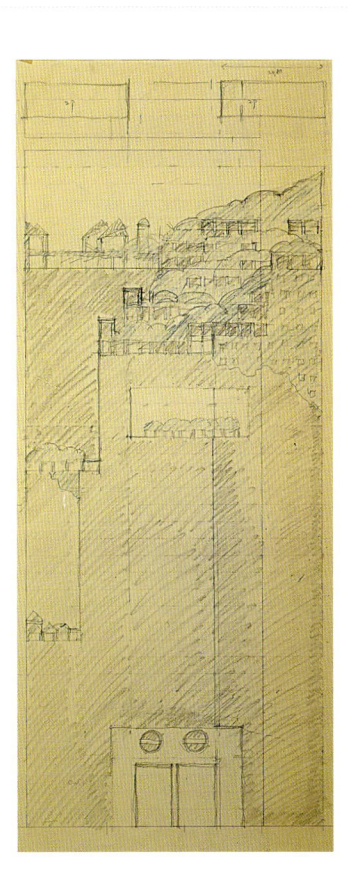

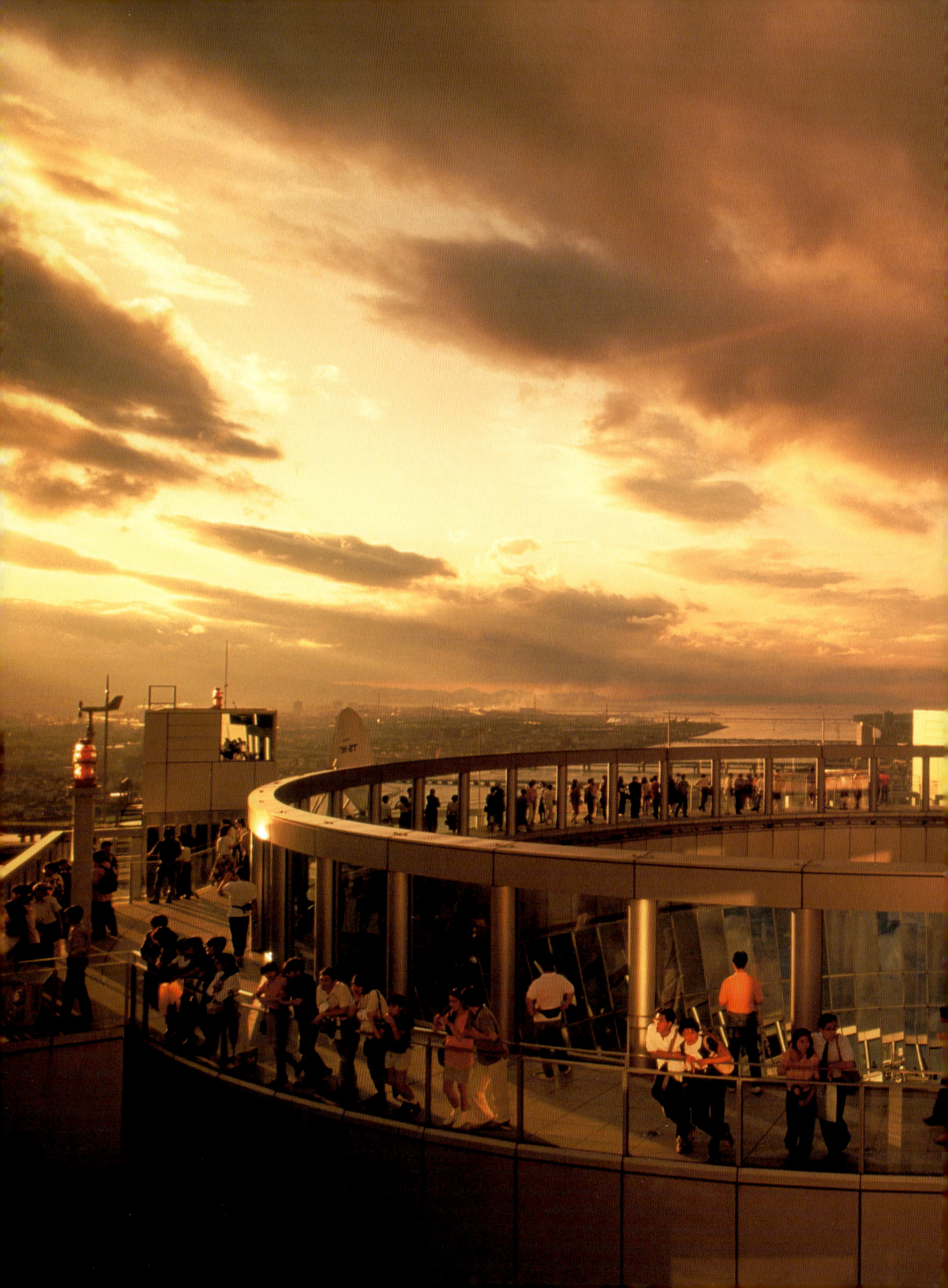

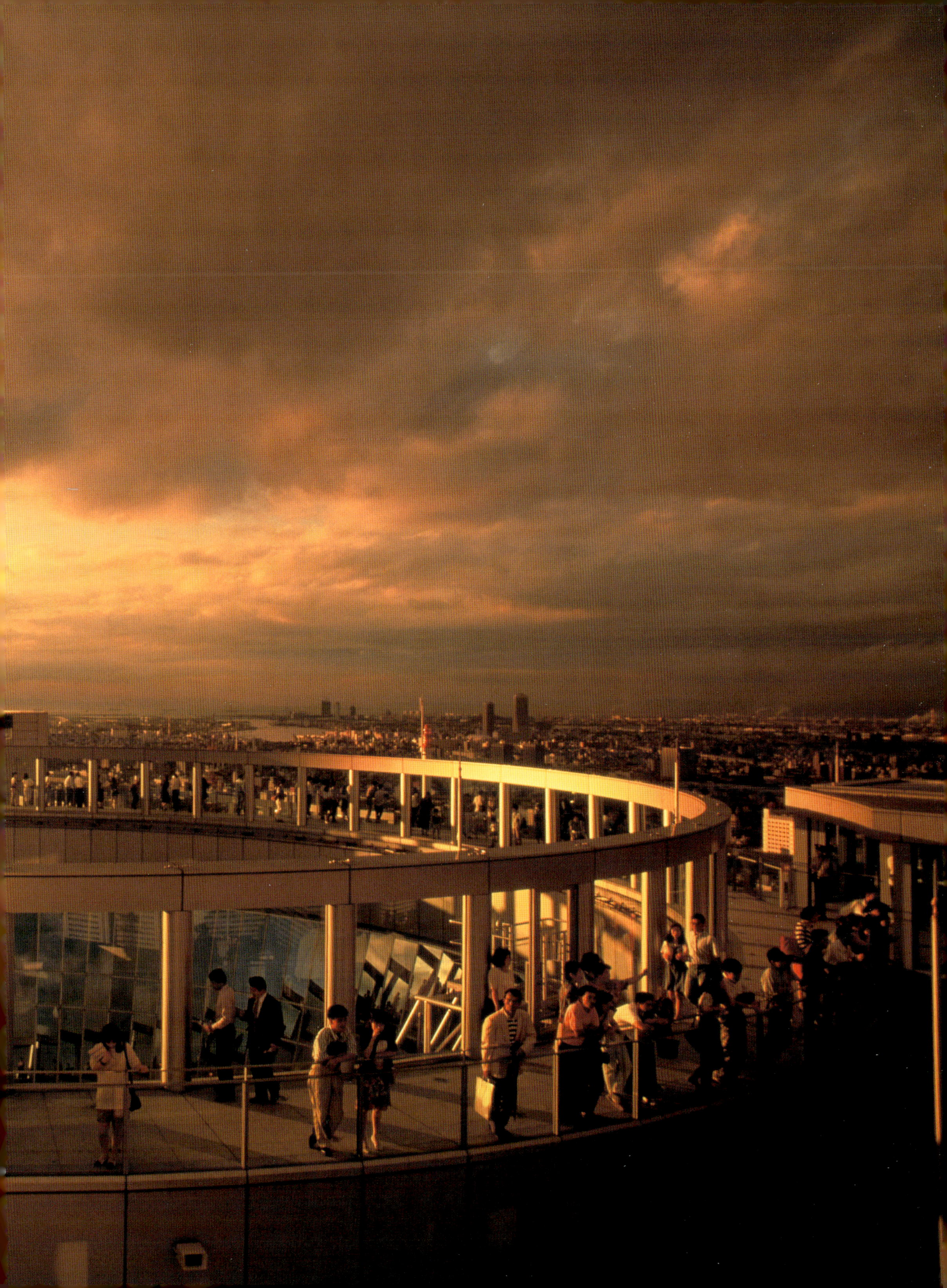

Hiroshi Hara

Umeda Sky Building

Opposite and above: Looking up towards the sky garden

Extra-Terrestrial Architecture – ETA (Project), 1992

The three significant aspects of ETA are the physical, the phenomenological and the social. I would first like to examine the physical aspects, best described as 'possible worlds'.

ETA is not an entirely new invention when one looks at the various space colonies that have been proposed to date, including different types of settlement on the lunar surface. In addition, it is difficult to distinguish clearly between a spaceship and an ETA. Despite these examples, however, to create a clear and practical vision of ETA we must introduce into our project certain guidelines based on technological hypotheses, even though they and their derivative predictions have no claim to authenticity. They are merely based on intuition, and cultivated by paying attention to the latest technological innovations.

One may feel that relying on intuition is dangerous. Investment in the field of space requires a great financial commitment. It is also necessary to make myriad technological breakthroughs before our intentions can be actualised. Still, if we contemplate the design of our current buildings and cities we can recognise that they have also depended on the strong guiding star of intuition and its power of creation.

Let us imagine a series of established technological hypotheses regarding architecture and outer space. Alluding to these conjectures, we can develop architectural renderings that I would call the images of the possible world. The possible world consists of hypotheses, mentioned above, and an agenda or programme. The combinations of hypotheses and programmes are manifold, resulting in a large number of possible worlds.

The phenomenological aspect is best described as 'space scenery'. Before discussing this aspect, though, I must first elaborate on the hypotheses introduced above. There have been many projects envisioning the future in the fields of urban and architectural design, including numerous proposals for utopia. These prophecies were also based on certain hypotheses extending beyond the technological achievements of the age. In time, some hypotheses were partially, and in rare cases, entirely, realised. There are many hypotheses still waiting to be realised, such as the vision of the mid-air city.

The reason we remember such projects is apparently not based on their realisation, for only parts of them came into being.

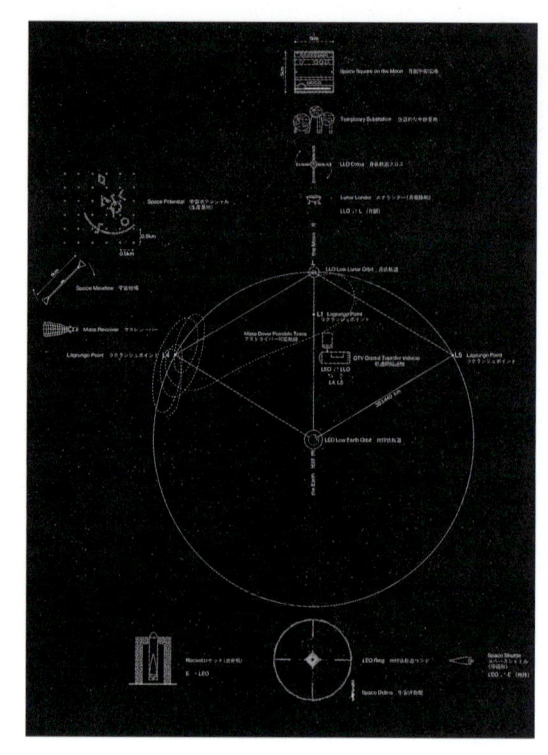

Formation of E.T.A. (Extra-Terrestrial Architecture)

Low Earth Orbit LEO-Ring seen from the interconnected super-skyscraper

Opposite Inset: Return to the Earth (Sky City and LEO-Ring)

Opposite: LEO-Ring and the surface of the Earth

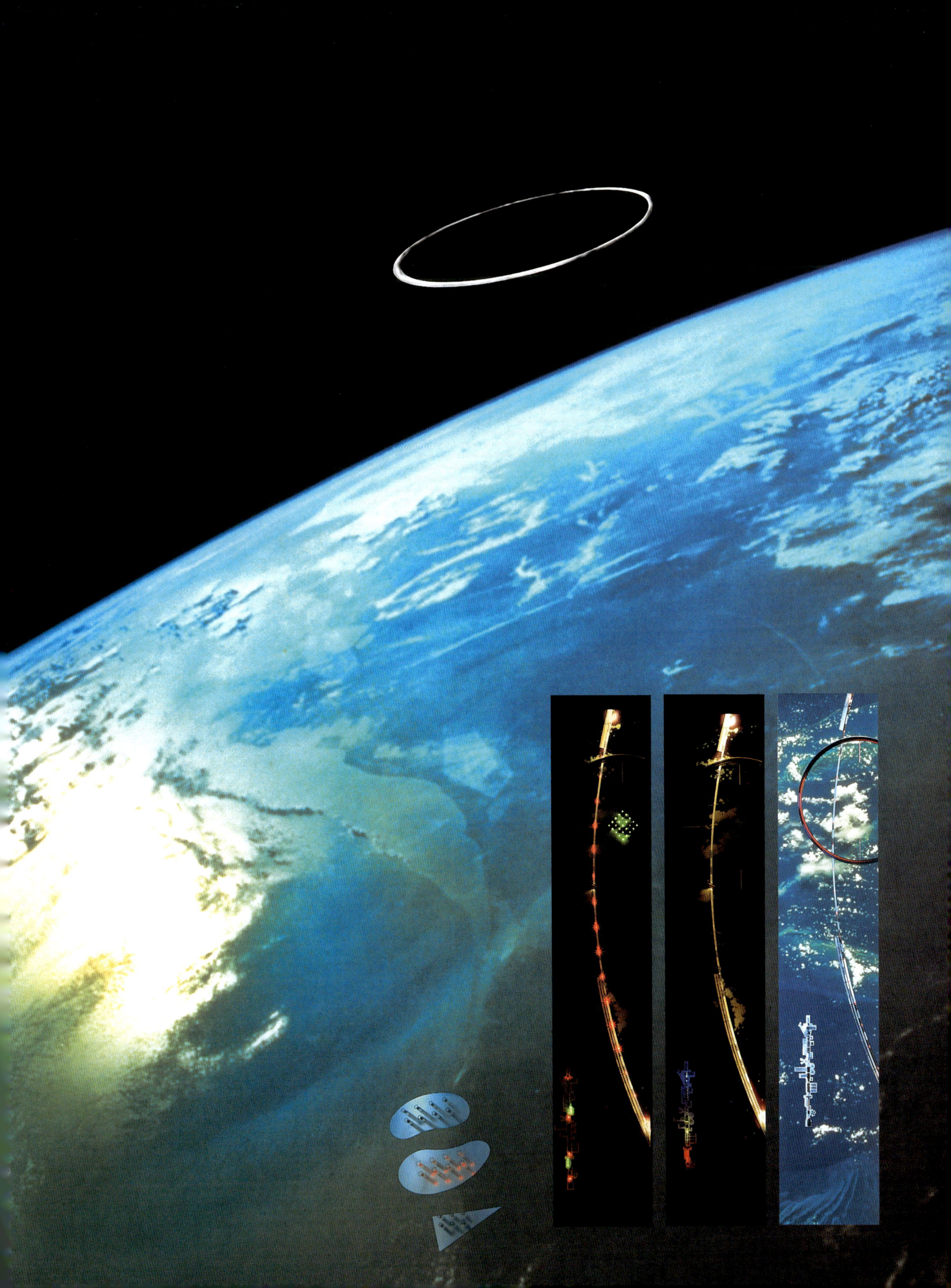

Hiroshi Hara

Space Square on the Moon

Underground habitat in Space Square on the Moon

Extra-Terrestrial Architecture

Right: Space Square on the Moon

Space Garden: windmills rotated by photoelectricity

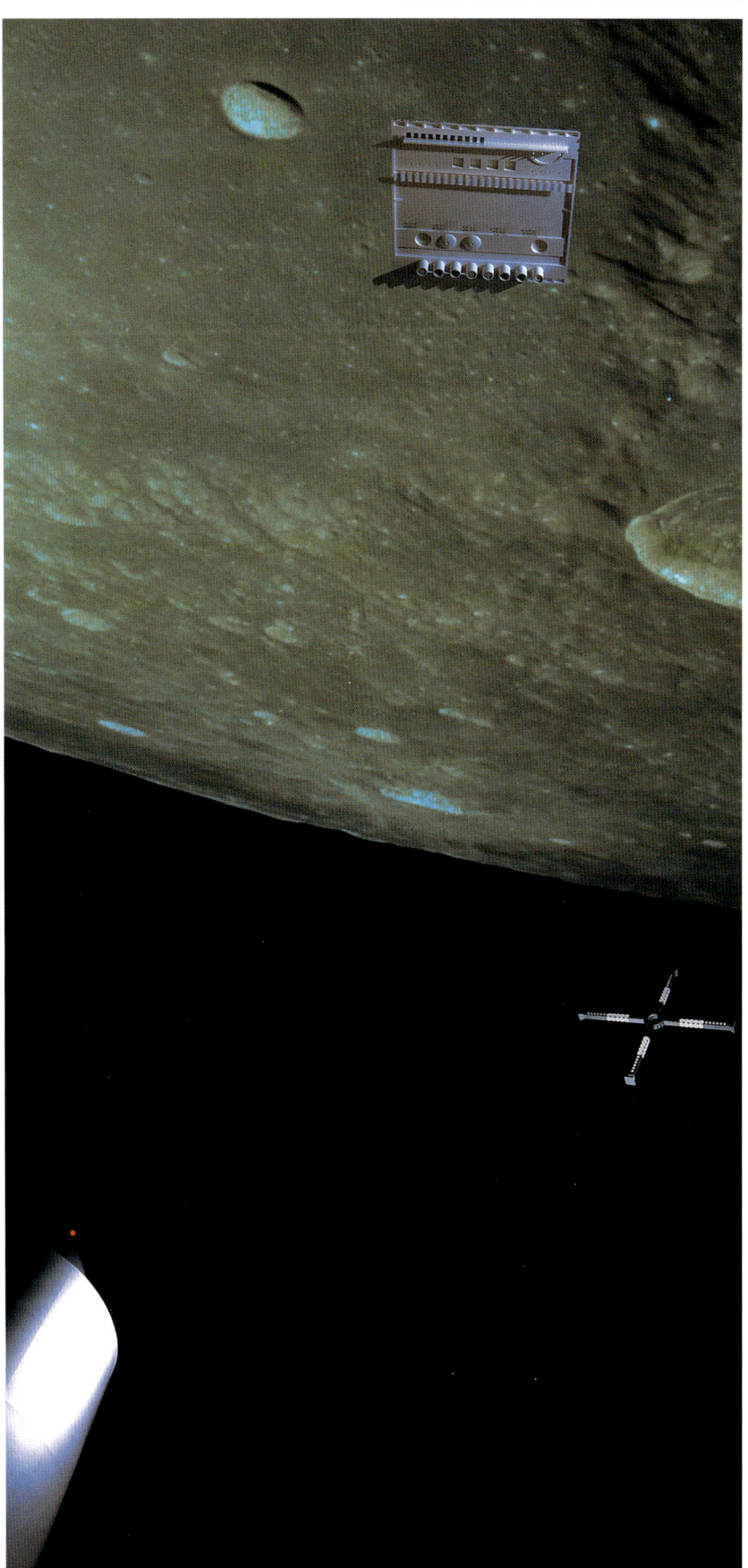

149

Hiroshi Hara

I believe it is not the reality that impresses people, but the vision or the scenery created by the fascinating power of the projects. This brings to mind the film *2001: A Space Odyssey* based on a novel by Arthur C Clarke and directed by Stanley Kubrick in 1968, or *The High Frontier* written by Dr G O'Neill in the same year. When designing ETA, I tried to transcend the images developed by such predecessors, yet I considered them to be as important as the technological fundamentals.

I would also like to suggest that one of the most crucial characteristics of an attractive project is that it should possess a monumentality based on fantasy. This fantasy has two roots, the first being the scientific-technological and the second the cultural-artistic. I would call the images generated by this fantasy 'space scenery'.

To date, various pioneering projects have shown us space scenery never seen before. We have been presented with a newly discovered view of nature. However, this view has often included the silhouette of a spaceship landing on the Moon or a human figure in a space suit. In other words, space scenery has, to a certain extent, been created artificially. The ETA project aims to transcend the design of this space scenery.

Previously, I have written papers on the perception of 'world scenery'. If the 'world' corresponds to our consciousness and 'nature' corresponds to an existence outside our consciousness, then space scenery can include and transcend world scenery.

Finally, we come to the social aspect. The space scenery described above involves humans. However, technological hypotheses may alter the depiction of this fantasy. The choice is so diverse that, in extreme instances, humans become redundant in space, their function being replaced by a troop of machines sent into orbit to work for us.

ETA does not exclude the possibility of functioning without humans. However, it is designed to form scenery that acts as a frame for people visiting outer space; it provides a place for them to interact in space. When two people meet in space, they share the same space scenery. This scenery is framed and formed by ETA. In this sense, ETA serves as a device that defines human relationships in space. Its aim is to depict the form of a future society in space. *HH*

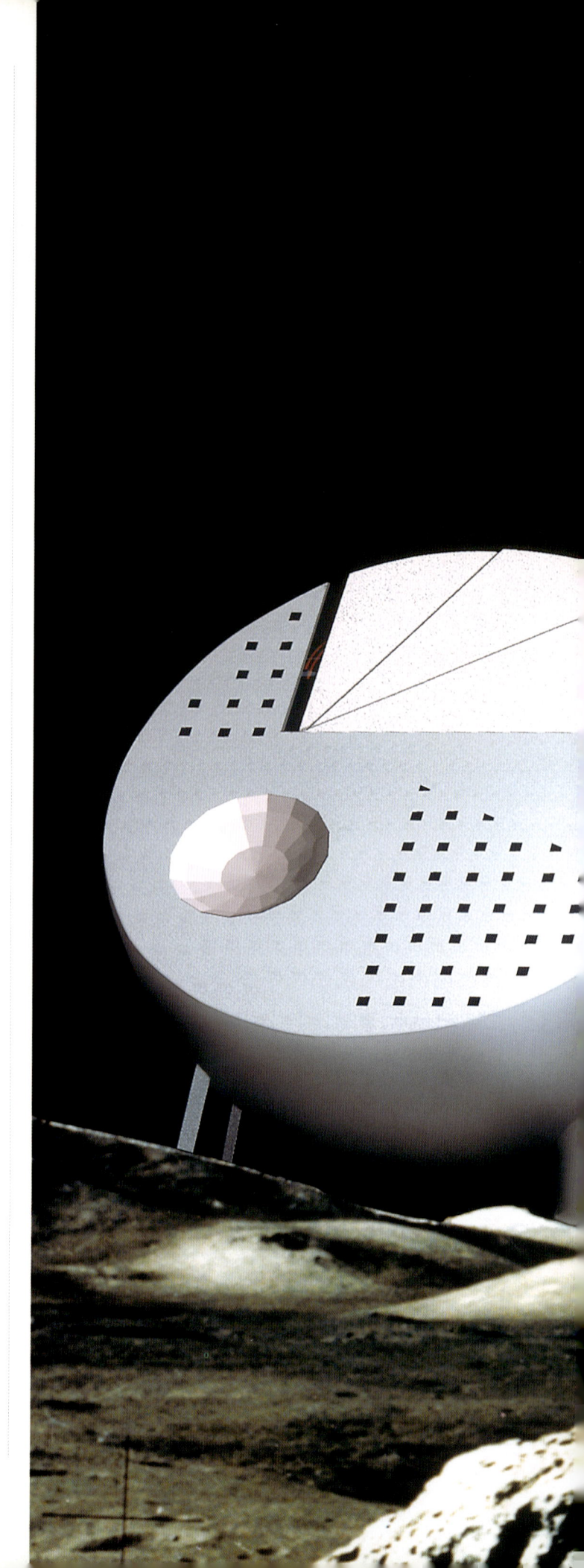

Temporary substation

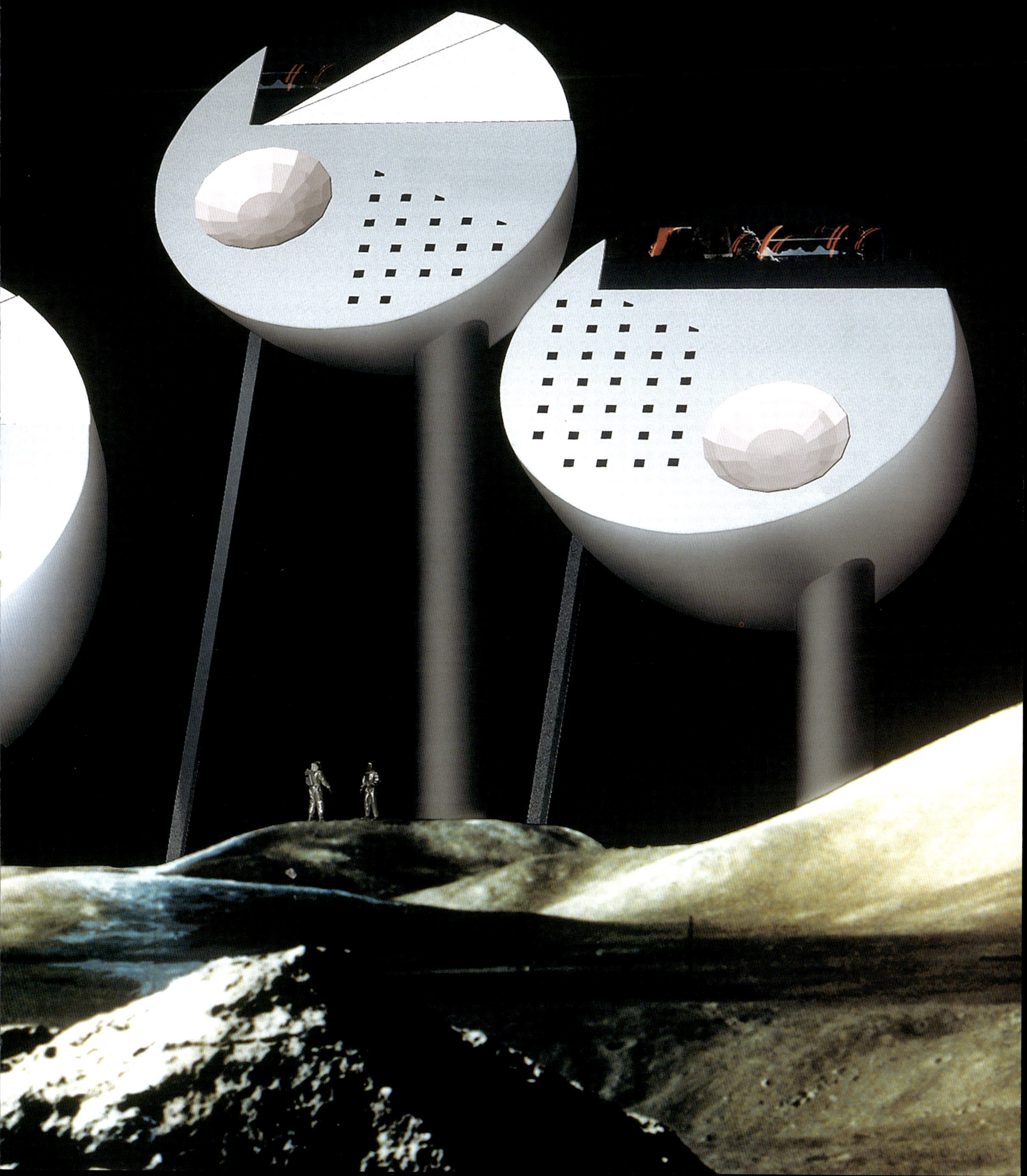

Four Cube-Houses, Tokyo, Nagasaki and Wakayama Prefectures, 1998

This relatively recent series consists of four houses that are dispersed throughout the country: one is in Machida, Tokyo, another in Wakayama Prefecture, and the other two in Nagasaki Prefecture. Each is a small, self-sufficient building; had they been built close to one another, they would still be independent. From a planning perspective, they are what I call 'floating elements'. That is, they are like an apartment block whose discrete units have been scattered.

I initially conceived the project as a series of five buildings. Each was to be in the form of a tower, with a cube placed on top of a platform. The dimensions of each cube are based on a mathematical progression with a common difference of 60 centimetres. Of the five cubes, four were built.

When I first began to practise architecture, my work consisted of small buildings such as houses. Gradually, the buildings became bigger, until eventually I was designing structures such as the Umeda Sky Building and Kyoto Station. I explored the design of even larger buildings, imagining a 500 cubic-metre volume and ETA (Extra-Terrestrial Architecture). In architecture, however, bigger is not necessarily better.

In the 1840s, the writer Edgar Allan Poe determined to

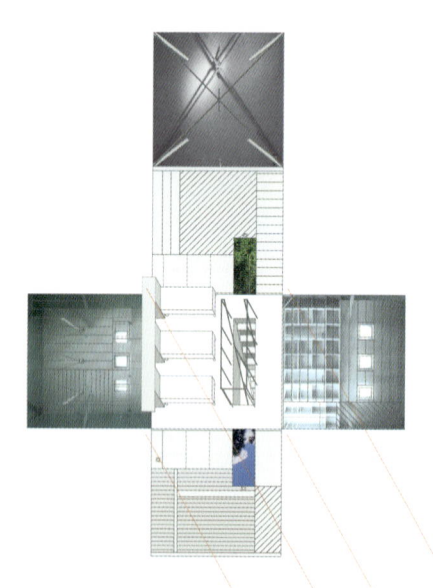

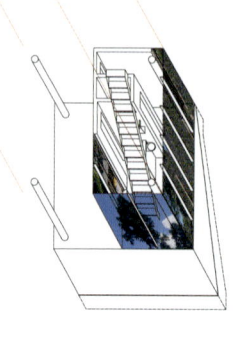

$(5200mm)^3$: the east facade; Nagasaki Prefecture

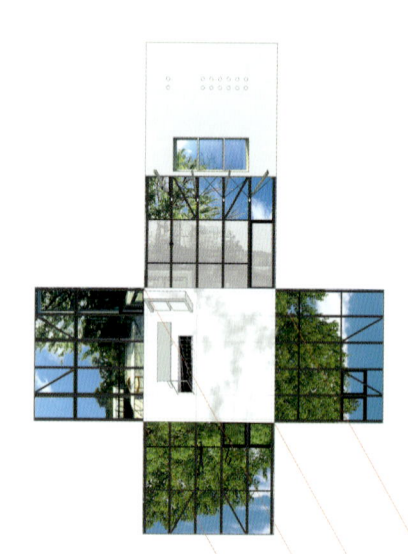

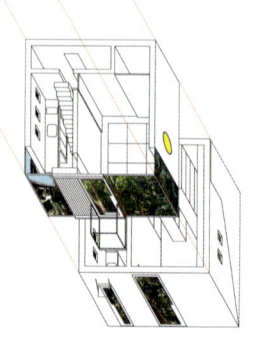

$(5800mm)^3$: the north facade; Machida, Tokyo

Four Cube-Houses

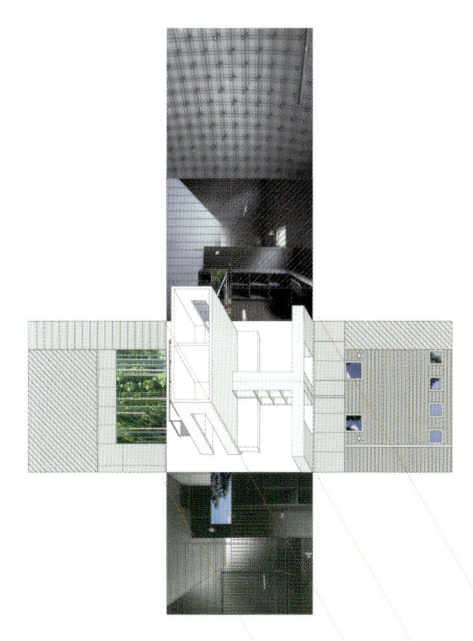

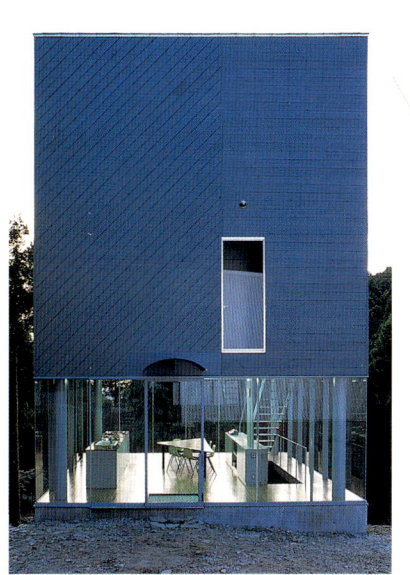

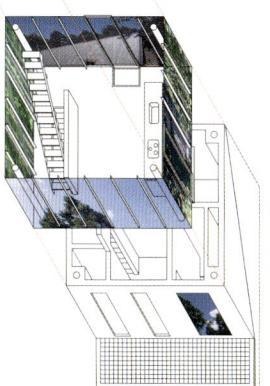

$(6400mm)^3$: the east facade; Nagasaki Prefecture

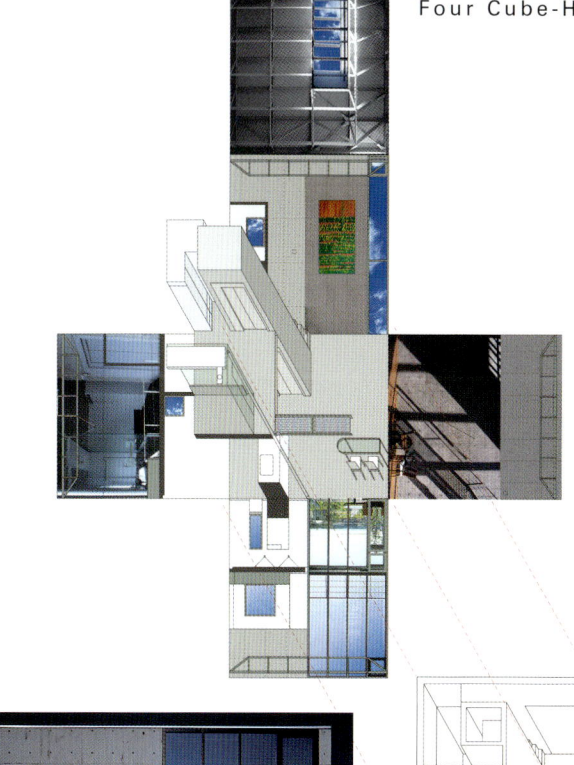

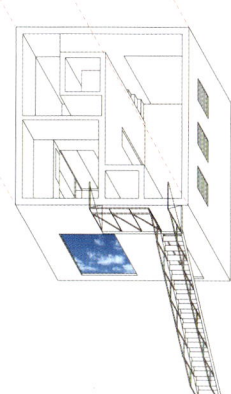

$(7000mm)^3$: the east facade; Wakayama Prefecture

create an immortal masterpiece of poetry. In order to ascertain the best way to do this, he studied famous poems from different periods of history and different cultures. He concluded that the ideal length should be 100 lines, which was approximately the length of his celebrated work, *The Raven* (1845). This was just one element in a wider methodology that includes a number of intriguing stratagems. Art has often been based on a similar methodology. In the case of architecture, acknowledged masterpieces have been, more often than not, small-scale works. I believe all architecture is an extension of the house.

In designing this series of houses, I asked myself the following question: if and when we manage to develop a computerised architect – let us call it ARCHITECT – what would its designs be like? I tried designing as ARCHITECT might, assuming for the purpose of the exercise that it had been programmed with the tenets of modern architecture. I conjectured that an elementary computer program would probably use as a reference three modern architectural models: glass boxes, experimental California houses and the aesthetic of collage, though examples of works based on the collage aesthetic are few. Each time I designed, I ended up going over budget. I could almost hear ARCHITECT snickering at my predicament.

One of the aims of this series was to make the house a small attractor in the city. In the near future, I will try in a different series of houses to design discrete attractors that are voids.

Overleaf: View from the southwest. Parents' house (left) and children's house (right), in Nagasaki Prefecture

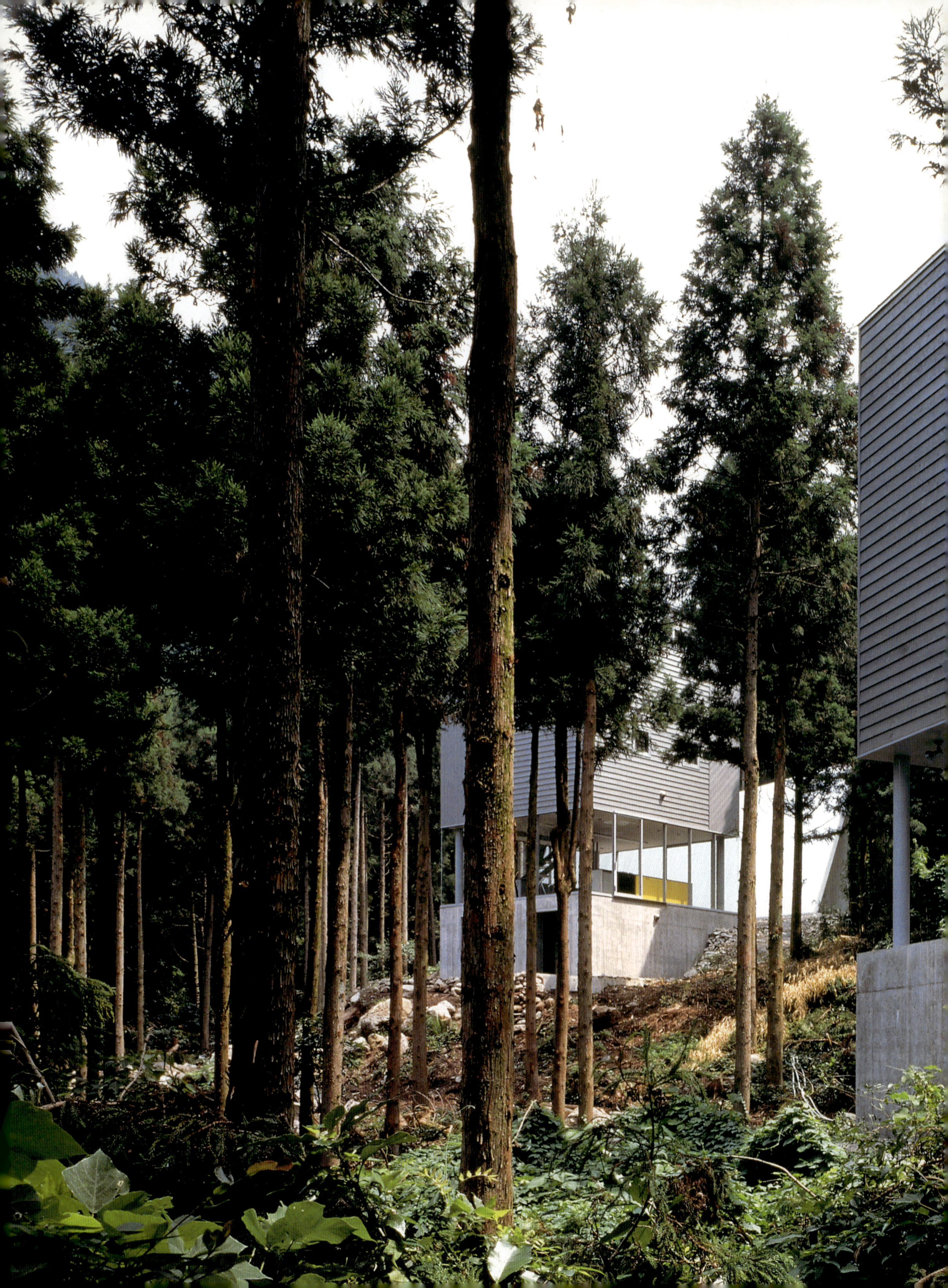

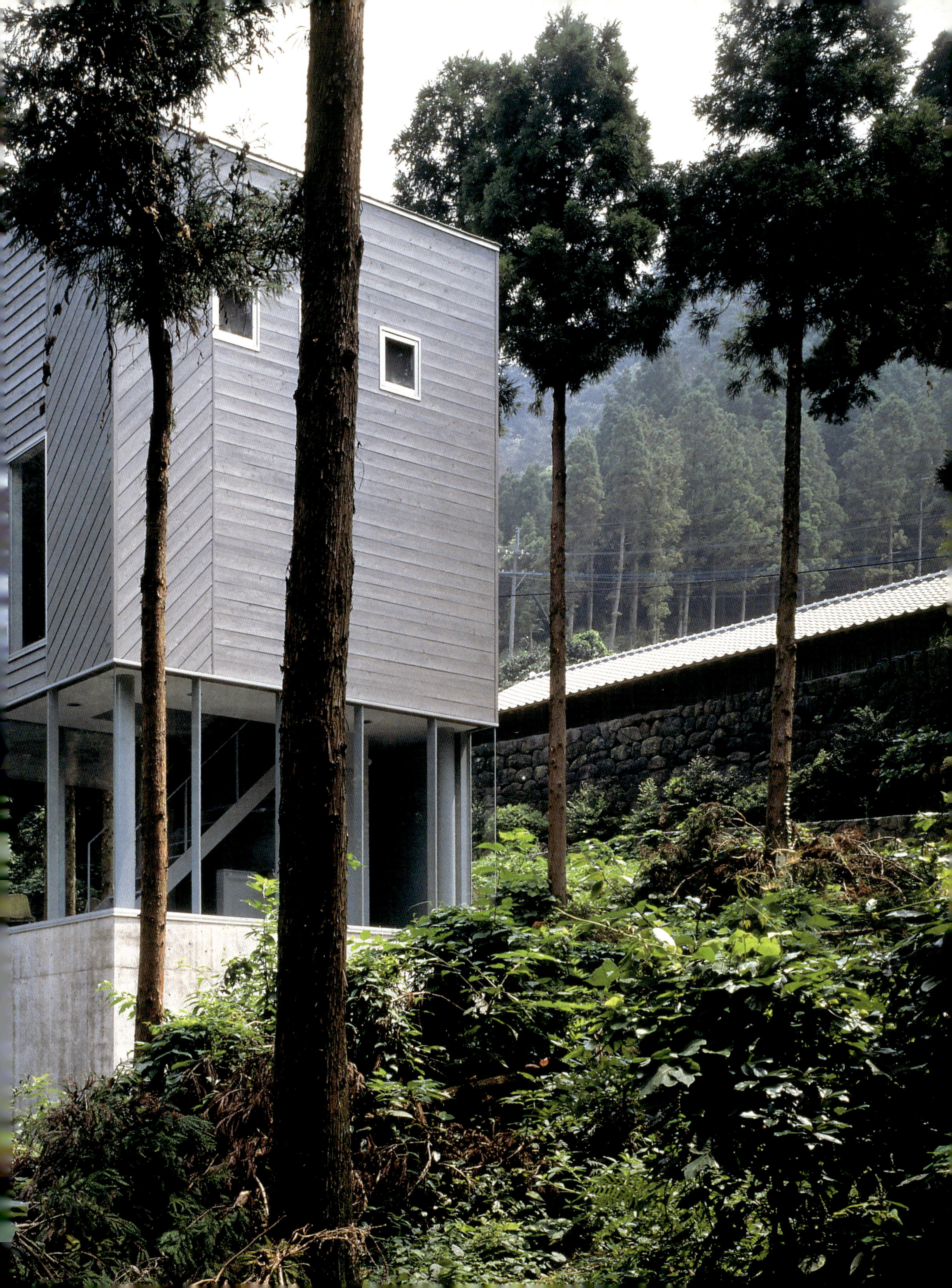

Hiroshi Hara

Four Cube-Houses

Opposite: Interior of the second floor of the parents' house in Nagasaki Prefecture

Interior views of the first floor

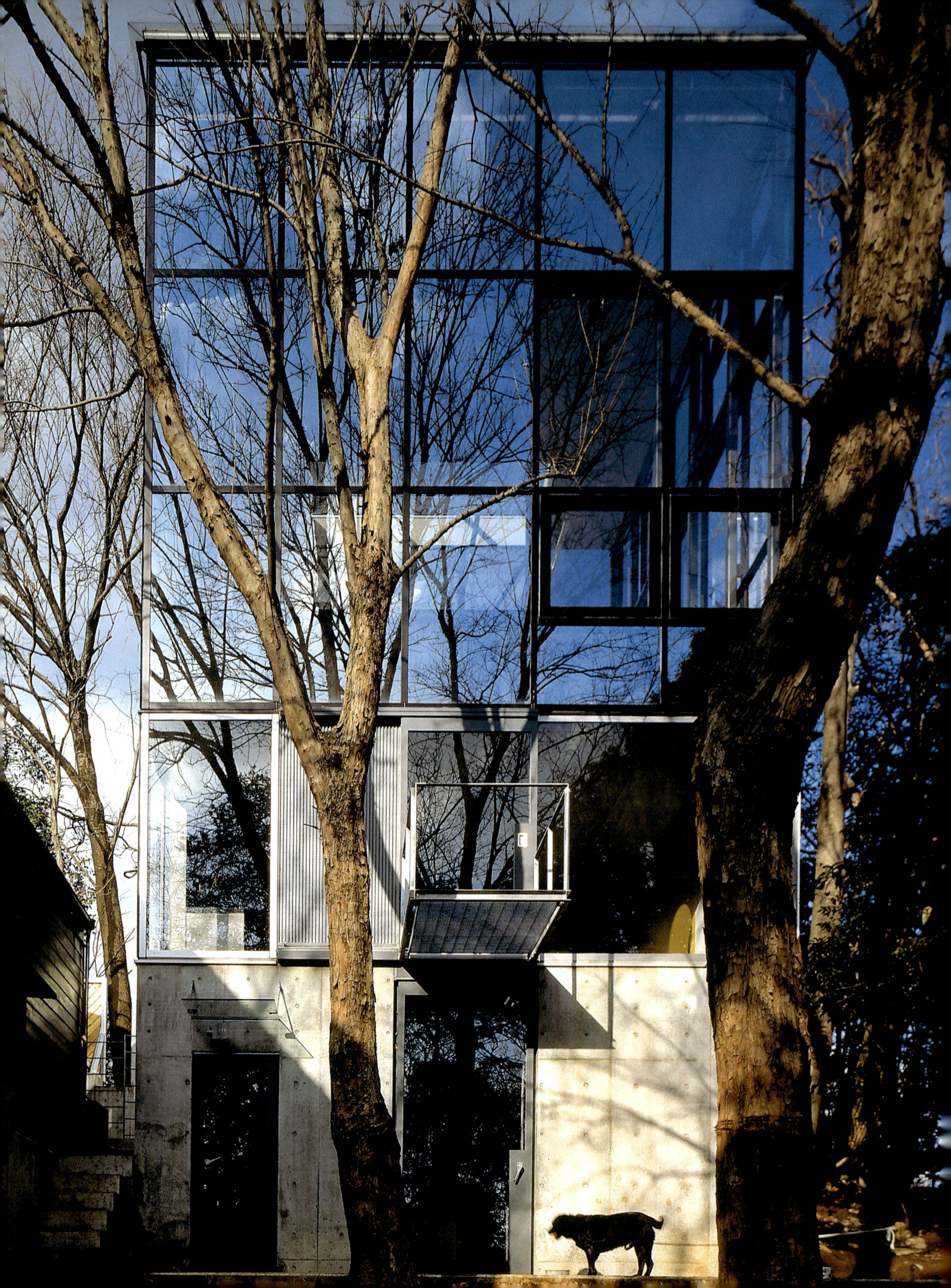

Four Cube-Houses

General views from the south in Machid, Tokyo

Upward view of the living room in Wakayama Prefecture

Four Cube-Houses

General view of the northeast in Wakayama Prefecture

5. Attractors / Semiotic Field
by Hiroshi Hara

As we have seen, to say that architecture involves not things but events is tantamount to saying that architecture is not things but space. Obviously, architecture has always been understood in terms of elements such as floors, walls and roofs. So as long as buildings are made of things, these elements will be necessary words in the architectural language system. However, the space created using such vocabulary is different from a mere set of things. For instance, the same building allows different phenomena to occur during the day and at night, and one cannot explain this difference in terms of things.

Ever since my doctoral thesis, I have been aware of this discrepancy between things and space, but I was not satisfied with the outcome of my research at this time. Later, when I undertook my urban research trips throughout the world, I applied it to understanding various phenomena in the concept of the field, which I have studied over the course of thirty years. People, buildings and cars are examples of urban activities that can be analysed as fields. Simple activities require scalar fields; the movement of people and cars requires vector fields.

From the viewpoint of the field, architecture comes closer to space than to things. Or to put it another way, things also form space in conjunction with phenomena such as temperature and light. When sunlight enters a room and the space becomes brighter, pillars and windows appear as space. This idea is based on phenomenology. In order to handle brightness and temperature, the best method is to consider things as symbols.

The concept of the field has been fully developed in physics. So, if things can be handled as fields, then space would be described overall as a field. Though many problems remain, I would like to call the domain in which symbols appear as phenomena the 'semiotic field'. Hence space is an overlay of the physical field and the semiotic field.

Just as phenomenology has suffered from both subjectivity and empiricism, so semiotics is also an area of some confusion. To some semioticians, the 'symbol' is an omnipotent concept that can describe any experience and, from this point of view, brightness and temperature are also symbols. In order to unify things and space, symbols must be defined from a limited, elementary level.

On normal maps, topography is illustrated by contours. Maps also describe the state of the land surface and usage. If the map is separated into areas using colours, then it becomes a semiotic field of continuum. But since this type of semiotic field is difficult to handle, we will delay consideration of it for the moment. On a map, one also finds symbols of landmarks and sights. These are mutually distant. Thus they can be described as the semiotic field of individuums.

Talking in terms of fields and semiotic fields is another attempt to give architecture a language system beyond that of floors, walls and roofs. But in this preliminary stage we can avoid expressing floors, walls and roofs as symbols, and concentrate on a semiotic field of a set of individuums, or a semiotic field of discrete assembly. These symbols, which appear mutually distant, can be called 'attractors'.

An attractor is a type of singular point that is discussed in geometry. As seen in the illustration, a point at the bottom of a pit gathering rain is an attractor. Conversely, a summit that diffuses rain is a 'repeller'. Architecturally considered, attractors are landmarks that attract people's attention, whereas repellers are exemplified by buildings such as fortresses, built in order to disperse the enemy. For instance, Yemen's houses, famous for their highrise appearance, are built on top of a hill to avoid nomad attacks. Churches, shrines and sightseeing spots represented on maps are examples of attractors. In today's city, repellers are rarely seen and almost all singular points are attractors.

I have submitted proposals to urban design competitions in Cologne (1987–8), Montreal (1990) and Majorca (1990). Recently I took part in urban planning workshops in Montevideo (1998–2000) and Cordoba (2000), Argentina. The concept of the semiotic field had become clear during the Majorca project, although I always had it in mind in my previous projects. All of these projects grasp the city historically and environmentally then develop a certain area within it.

Attractors / Semiotic Field

In other words, I considered the entire city as a semiotic field and added attractors in specific areas to revitalise it. Cologne's Media Park, Majorca's Parc BIT and Cordoba's TELEPORT are information centres for the new media age. The facilities therefore had to function as attractors for the entire city. I filled the facilities with small attractors so that they become one large attractor.

The International City project for Montreal was intended to mediate the famous old city and the rapidly expanding modern city. The historical buildings that are preserved here and there function as small attractors. The project consists of five overlaid strata, which could also be considered attractors.

In the TELEPORT project in Cordoba, I used overlay, ie, multilayered structure, and refurbished some currently unused buildings as attractors. Along the river, natural elements and electronic devices are interlaced to form attractors called 'piers'.

These are large-scale semiotic fields that extend to the entire city. Small semiotic fields realised in a single building can be exemplified by Kyoto Station and Miyagi Prefectural Library (both 1997).

Kyoto Station is a 230,000 square-metre urban complex consisting of a department store, hotel, theatre, parking lots and station. The station occupies a mere fifth of the total area. Keeping this programme in mind, in order to make the entire building a station, I placed the concourse at the bottom of the V-shaped valley so that people would have an overall view of the building while standing within it. One side of this valley connects to the department store and the other to the hotel. Escalators climb up each of the slopes. The slope on the hotel side is somewhat quiet in comparison to that on the department store side, which has a large staircase and is more lively.

Taking valley topography as the base, I placed on it various attractors of different sizes to form a discrete semiotic field. Smaller attractors are the lighting towers and sign boards. Medium-sized attractors are the equipment towers that contain air conditioners, several types of follies and radiant art displays by Roy Lichtenstein, Robert Longo,

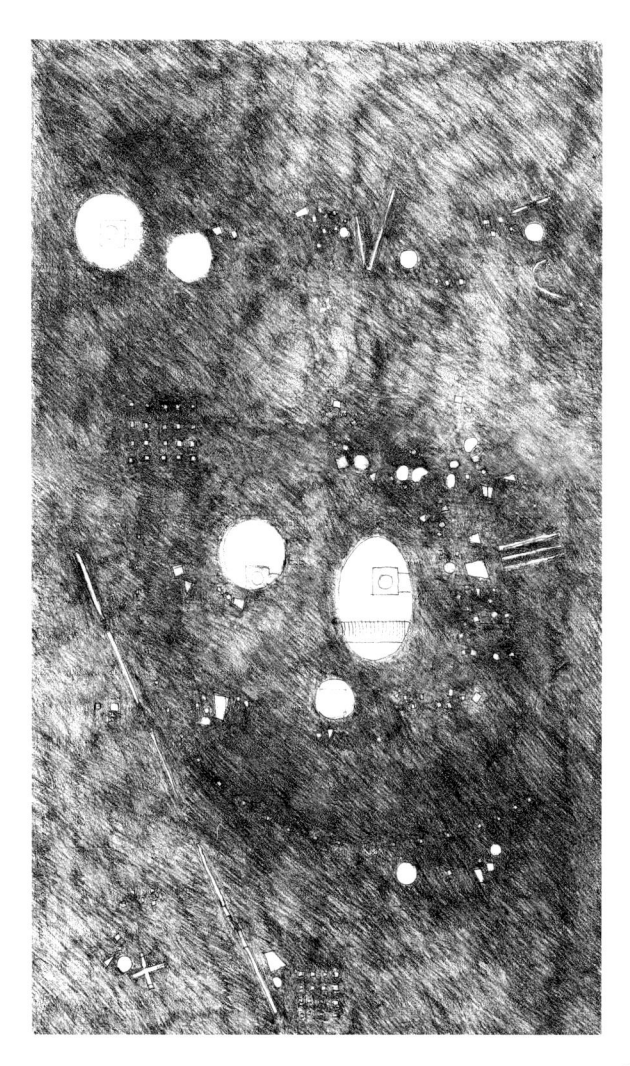

Kyoto Station Building: the ceiling of Grand Theatre in Theatre 1200

Joseph Kosuth, José de Guimarães and Tadanori Yokoo. Bridges and mid-air paths form floating attractors, though they cannot be called discrete. The truss roof is a large attractor.

However, the best discrete attractors are the people who climb the valley. People enjoy watching each other from various viewpoints. The valley topography urges them to do so, like actors on a stage.

In front of Kyoto Station, instead of a pedestrian plaza, there is an area for taxis and buses. The valley/semiotic field therefore becomes a three-dimensional plaza within the building itself. Every day, many events take place in the valley. People can choose from several paths, according to their individual preference, in order to wander around the valley. For instance, in the department store, an escalator is situated parallel to that along the valley, while under the truss roof is a mid-air path that crosses the valley and at the same time commands the city of Kyoto. Here, the attractors function as a means of navigation.

In the Miyagi Prefectural Library, I also had the semiotic field in mind, though the two buildings have contrasting programmes. Kyoto Station is an access point to the city visited by about 400,000 people per day. Miyagi Prefectural Library, on the other hand, is situated in a suburban area and is used daily by only 2,000 to 3,000 visitors. While the station is a multicultural venue, the library deals in high culture. Despite these differences, as long as the building has a public aspect then it should be designed as an extension of the city.

The site of the library consists of several valleys traversed by flowing streams. The library is designed as a bridge that crosses these valleys. The topographical plaza is an agora with stairs, situated within one of the valleys. With this plaza in the centre, the site is designed as a park, as far as possible preserving nature. Follies, paths and lighting relevant to a park are provided, designed by international artists. These artistic works function as attractors in the park. Other elements, such as walls, a valley spring, parking shelters, an entrance canopy, lighting towers and ventilation towers are also intended as

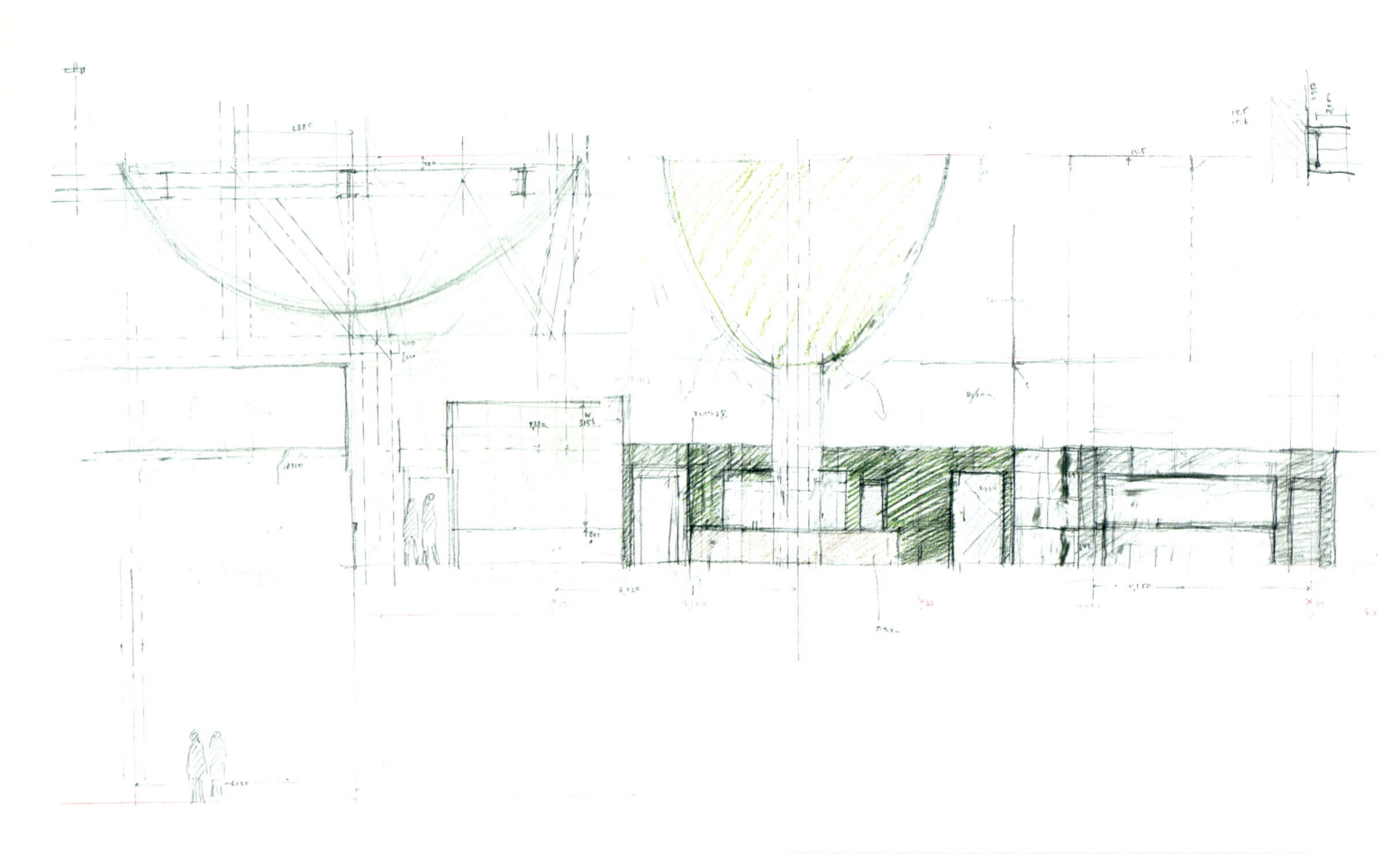

Sketch of Kyoto Station Complex: attractor

Attractors / Semiotic Field

attractors. These are discretely installed and function as small landmarks. I also designed the desks and bookshelves, items of furniture that function as small additional attractors within the interior space.

This open library is a mall type, designed like a street so that the atmosphere of the city enters directly. In the reading room, the design of the ceiling is especially important. Here, in contrast to the concave topographical plaza, the roof is convex – like a wave – in two spots, from which natural light enters. In several places the inner steel frames are exposed. These are attractors that allow visitors to locate their position in the building.

If one concentrates on the discrete semiotic field, putting the design effort into attractors, there is a tendency for floors, walls and ceilings to be left neutral. Floors are inevitable as bases but would be genuine semiotic fields only without walls and ceilings. The result is that architecture comes closer to the city.

Housing project: attractor as void

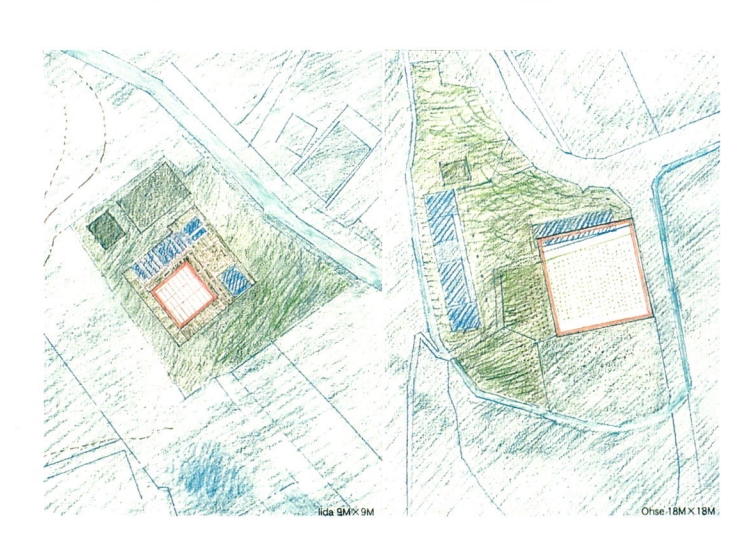

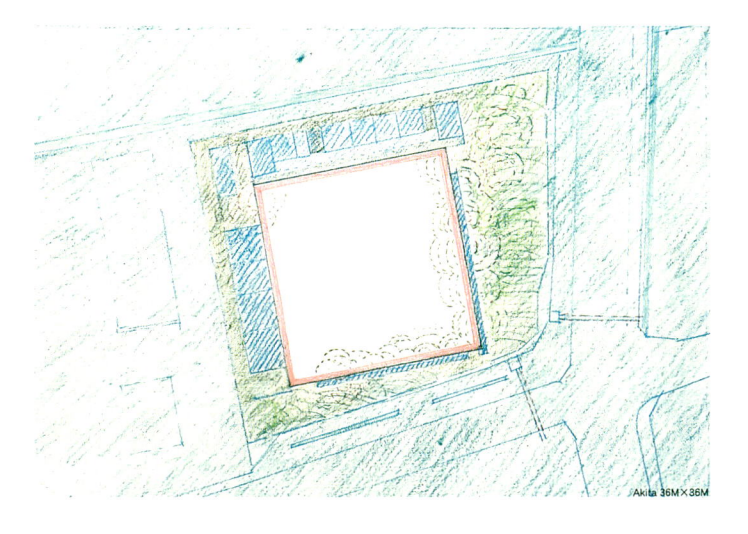

MediaPark Köln, Cologne, Germany, 1987–8

Our presentation team for the MediaPark consisted of Kei Minohara, Shimizu Corporation, NTT, and myself. Members of my research unit at Tokyo University completed the project team. We explored the kind of facilities that should be included in this new expression of electronics in the urban entity, in order to fit into the existing urban fabric.

After extensive research, we concluded that a ring, or circle, best represented contemporary media and electronics. We gave the idea physical form by designing buildings around a plaza. The plaza is a reinterpretation of a medieval town square as a centre for information exchange. The proposed 500-metre length of the plaza also derives from its medieval predecessors. The buildings surrounding the plaza are transposed from some of the more important streets of medieval Cologne. We fitted them into place through transformation operations such as reduction, division and articulation. Here, then, is yet another way of embedding fragments of the city into architecture.

The plaza itself is made into a stage set, which consists of underground wiring and equipment, as well as a cluster of multifunctional towers. It is a 'possible-world plaza'. By this I mean that when occasion demands, the plaza turns into a stage, unfolding tales via electronic media technology. On the other hand, the planning of this informational and entertainment area in relation to the actual urban fabric creates a static world. It is only when people are added to the formula that information is produced through human encounters. I was also greatly influenced in the design of this project by the German *Kammerspiel* film. *HH*

Plans

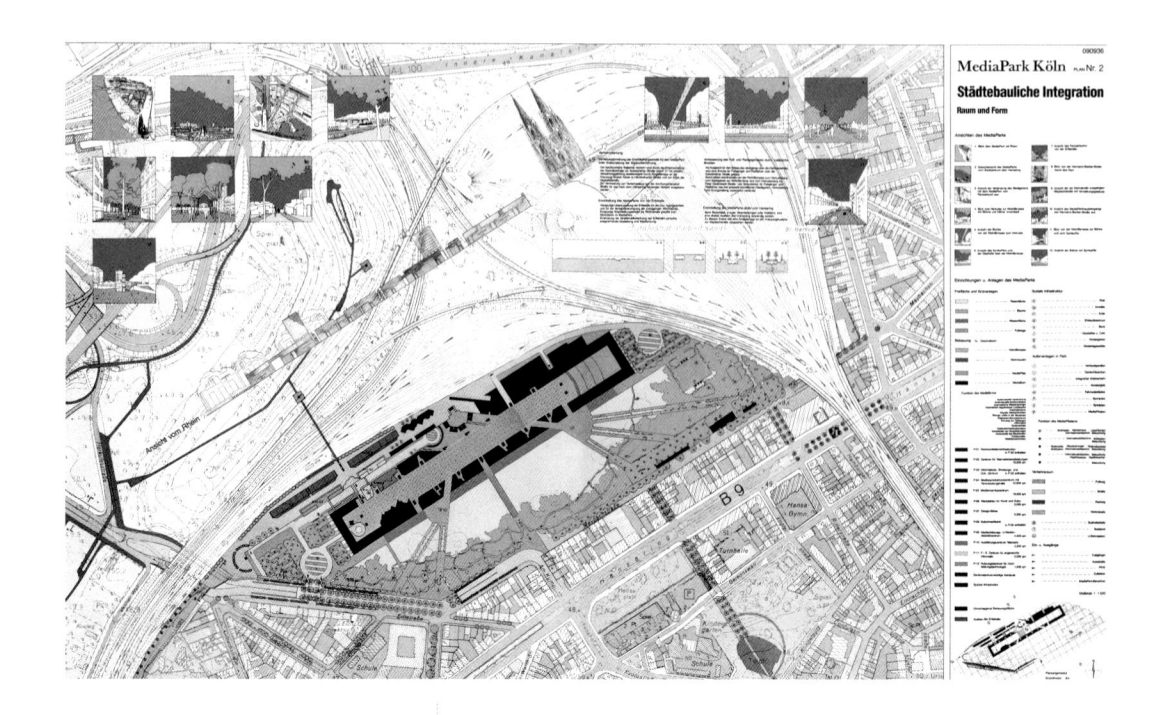

MediaPark

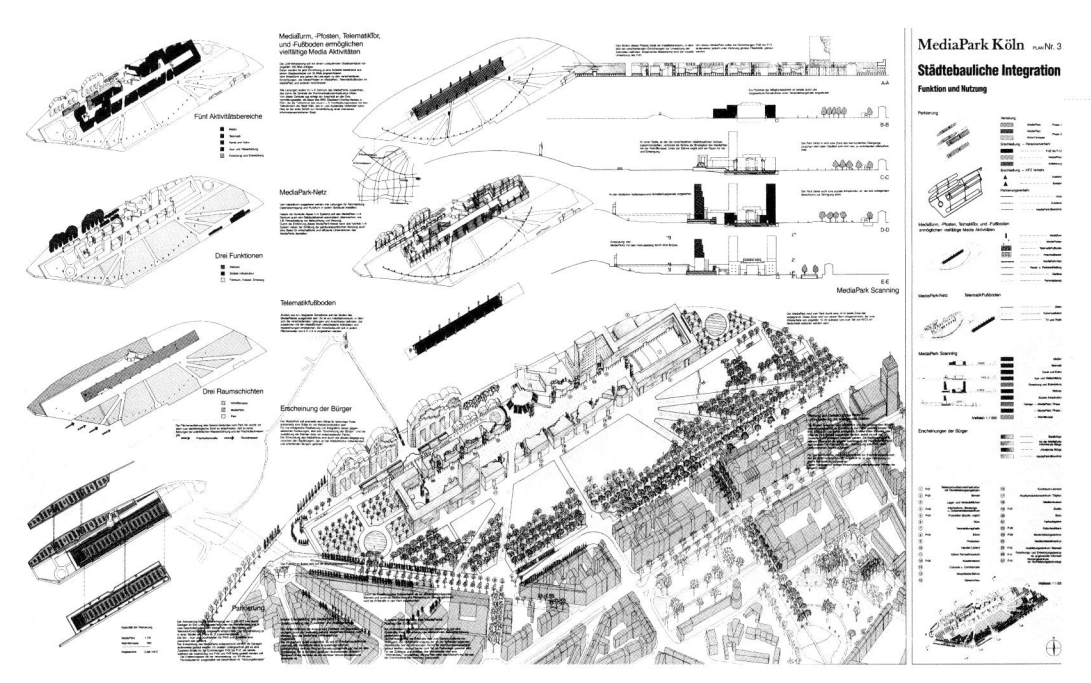

Model: east view

167

Hiroshi Hara

The International City, Montreal, 1990

To prepare the way for environmental improvement targeted for the year 2000, in 1990, Montreal held a series of city-planning related international design competitions. One of these, 'The International City, Montreal', was a redevelopment project for a location between the central commercial and business district and an area that was rich in history, divided by a large-scale expressway. The aim of the competition was to come up with an effective way of overcoming this physical separation and connecting the divided parts, and to propose a conceptual device that would give Montreal an identity as an international city. More concretely, the competition programme required office space and an international conference hall with a combined total floor area of 700,000 square metres and public open spaces in approximately 80,000 square metres of open land within the site. The competition participants were asked to provide not so much specific building forms as urban-planning principles or urban-design guidelines.

We presented the following four concepts. Each is a significant element for the spatial structure of the International City, Montreal:

a. Cross-shaped Plaza
b. Four Quarters
c. Urban Fabrics, Frames, Accommodators
d. Urban Strings and Floating Elements

The *frame* as an architectural element corresponds to the direction of the *cross-shaped plaza*. The *accommodators*, which have acute-angled triangular plans, are inserted between the various *frames*. The juxtaposition of the historic urban fabric and the *frame* is realised by the *accommodators*. Abundant but controlled *urban sceneries* will be generated by this operation.

The ten implicit *urban strings* are expressed by the arrangement of nodes. The nodes are named the *relais* – a variety of follies. The *urban string* is orthogonal to the *green belt* and parallel to the ICC office building. Pedestrian movements are induced by the *strings*.

The *floating elements* are placed around the *urban string*. They are arranged freely within the order of the *frame*. Wings, *shelters* and street furniture are examples. The giant *wings* belonging to the *floating elements* and the *symbolic frame* distinguish the urban silhouette of the International City from other cities. The smaller *frames* and other *floating elements* organise the human-scaled urban *sceneries*.

The existing city and the new multilayered urban scheme made up of the four spatial components above are considered to be the overall spatial structure of the international city. Plazas, greenbelts and various devices in the basic proposal suggested by this concept will help form views appropriate to an international city. *HH*

Photograph of Model

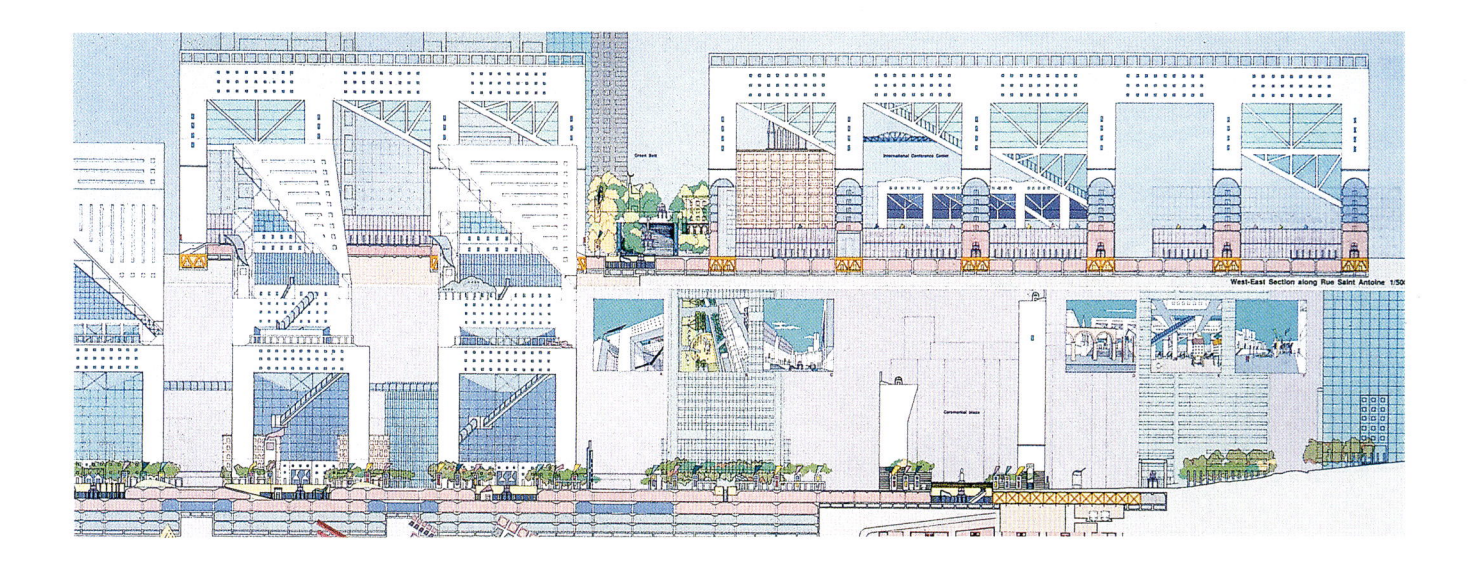

The International City

Hiroshi Hara

Parc BIT (Competition Entry), Majorca, 1990

This project was entered in a competition for an image and systematic framework for Parc BIT, which is being planned for Majorca. As the main island of the Balearics in the Mediterranean off Spain, Majorca is a major European tourist destination. The organiser of the competition, the prefectural government, intends to transform 150 hectares of land in a suburb of Palma in the middle of the island into a domain that will stimulate the development of new forms of communication.

We believe our proposal succeeds in being innovative in content while preserving nature. A flexible system is vital since the group of buildings has an urban scale and is to be constructed incrementally.

Our proposal is therefore based on a matrix organised around a land-use or building-use axis – that is, a functional axis – and an axis indicating the sequence of construction and differences among elements having the same uses – a transitional axis.

In the centre of the matrix is an 'electronics garden', a multi-stage plaza, as it were. Similar plazas will be constructed on other islands and enable new forms of communication between islands to develop.

The core of the project is a garden with a 'clopen' structure, (simultaneously open and closed). Thus the urban image is of a city with a garden having a new aspect, and the entire site forms the semiotic field of the garden. The project is arranged so that people will discover diverse pathways, and this makes possible the realisation of the semiotic field. The network of paths is based on the matrix created from functional-transitional axes. In traversing this matrix in diverse ways, people will make unexpected encounters. *HH*

Model photo

Parc BIT

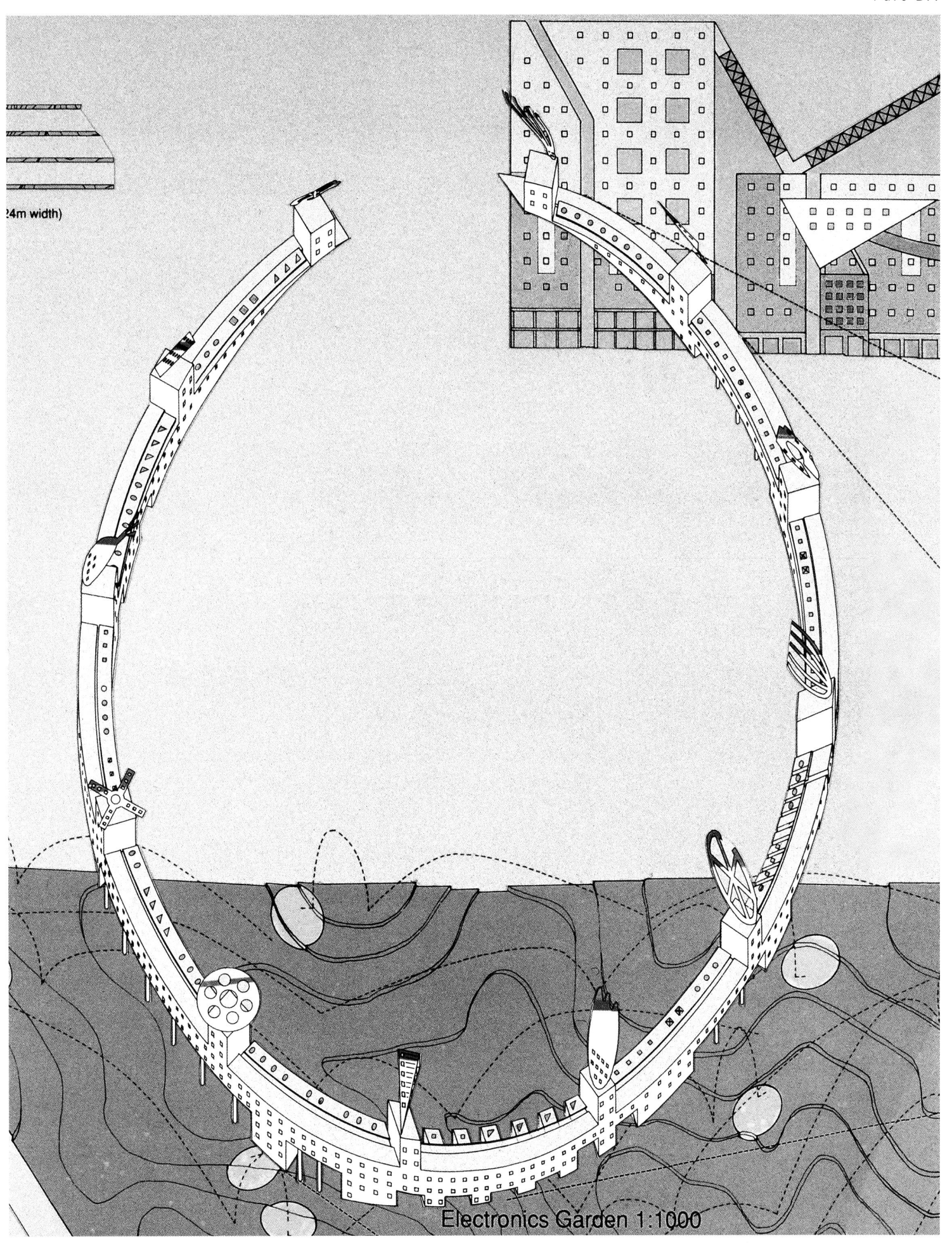

(24m width)

Electronics Garden 1:1000

Miyagi Prefectural Library, Sendai, 1997

In this project, I attempted to leave open as many options as possible for later additions and alterations, since the future state of the library was impossible to anticipate. I also tried to design the building in the shape of a bridge in order to preserve the original topography of the site. I was pleased by the way cylindrical rooms were combined at the top and bottom in my initial sketch, but at some point it occurred to me to make the cylindrical form wavy in order to provide a contrast with the topography. Creating this form with metal panels was difficult. After repeated trial and error, a method for twisting aluminium panels was developed and eventually adopted.

This building makes use of large quantities of glass. In the curtain wall in the entrance lobby, the horizontal force of an 11-metre-high glass surface is transmitted to a glass transom. The feature was developed by Toshihiko Kimura, the structural designer. In the conference room are three unique glass tables.

While a student, I received instruction from Professor Yoshichika Uchida, who is not only an excellent architect but a leading authority on building construction. He takes an intellectual interest in architectural details and often impressed on me the importance of detailing. In Japanese, the good fit of a detail is referred to as *osamari*. Not just a matter of the mechanism of the detail, *osamari* also involves such things as durability, ease of maintenance and elegance. It was from about the time I designed the Umeda Sky Building and Kyoto Station that I began to address Uchida's high standards with respect to details. This building type generates many details, particularly since I have a tendency to pursue a heterogeneous order. Trying to figure out the *osamari* of a detail is fascinating work for me, but to see to all these details personally entails an enormous amount of work, which requires the involvement of craftsmen and specialists.

I feel that on this building is successful with regard to its detailing. Architecture is not something one creates by oneself, yet there are aspects that can only be created in this way. The professional architect is faced constantly with such contradictions. *HH*

Exterior view of the entrance approach

Opposite: Exterior view of the north

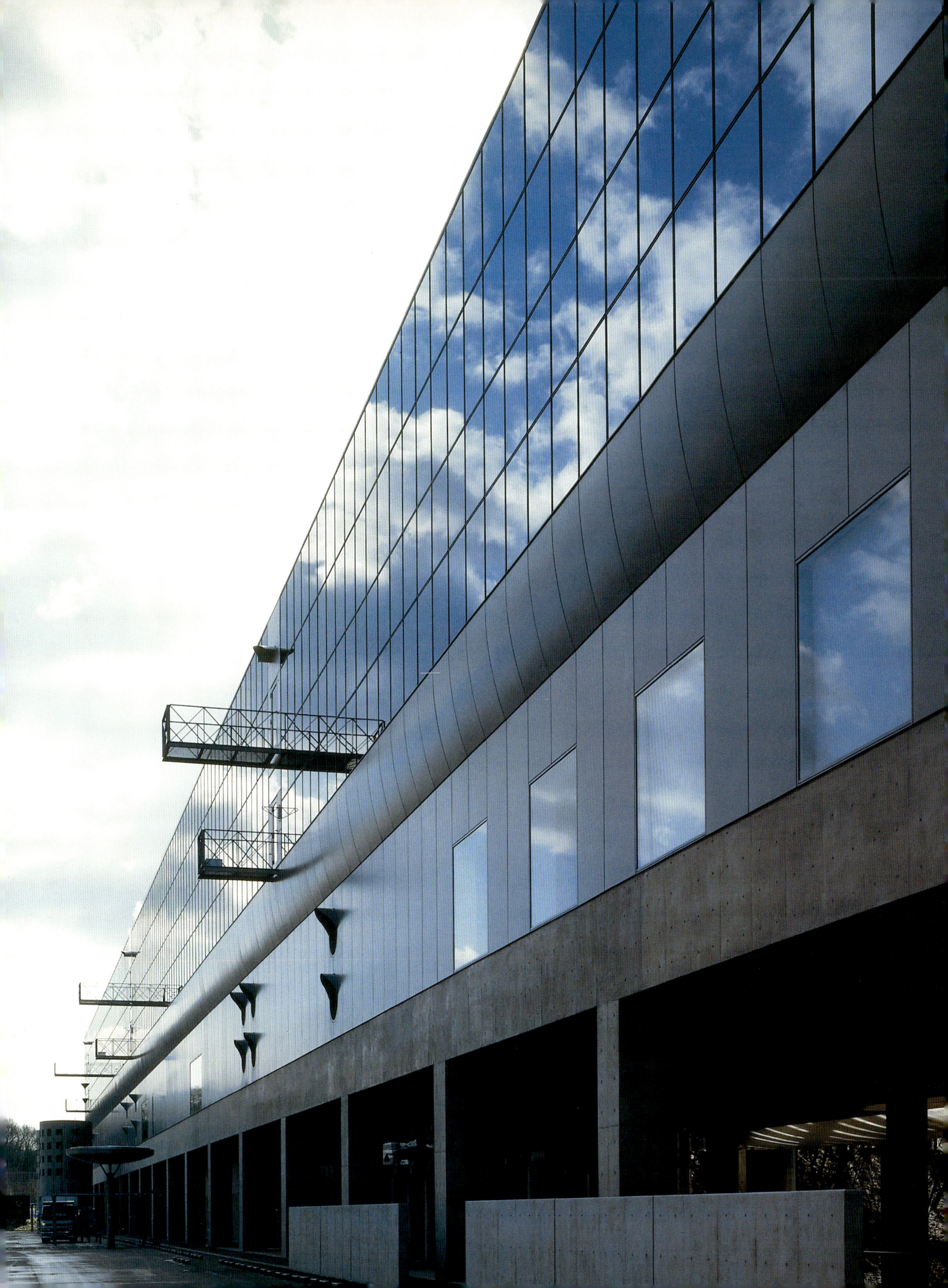

Exterior view of the north

Roof plan

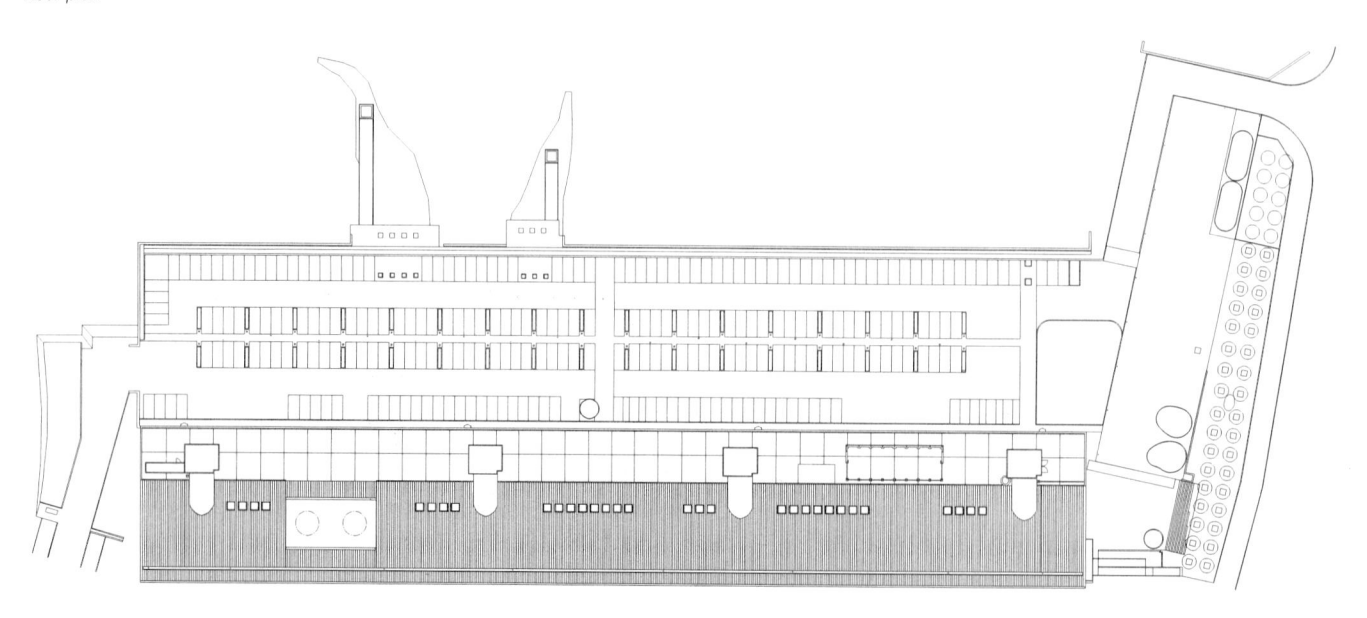

174

Aerial view

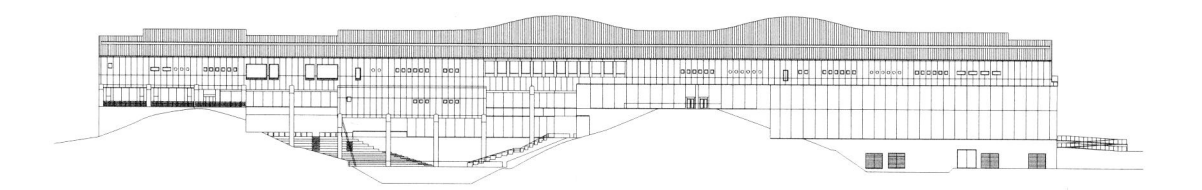

South elevation

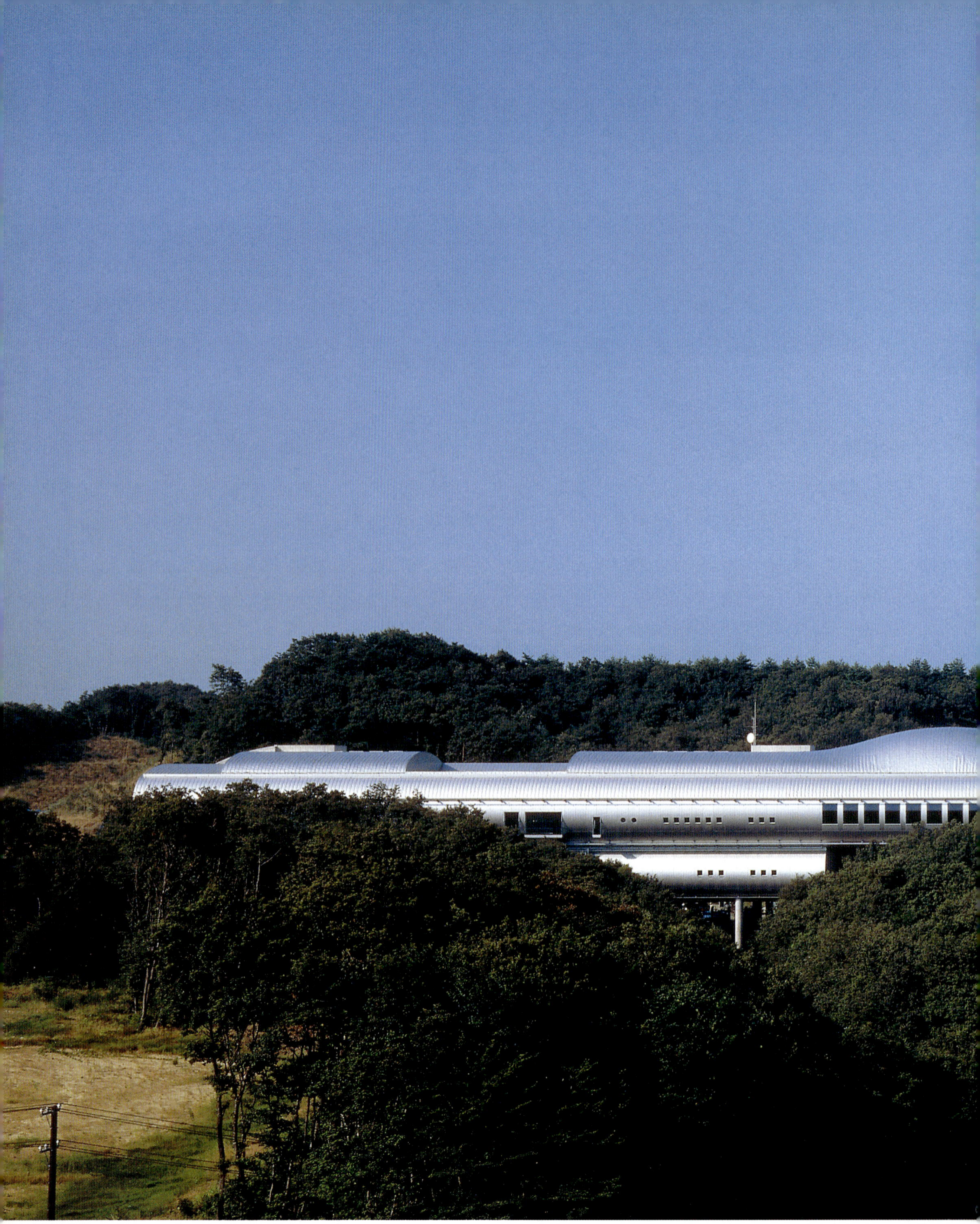

General view from the south. The building is set amid natural surroundings

Joseph Kosuth
Twice Defined

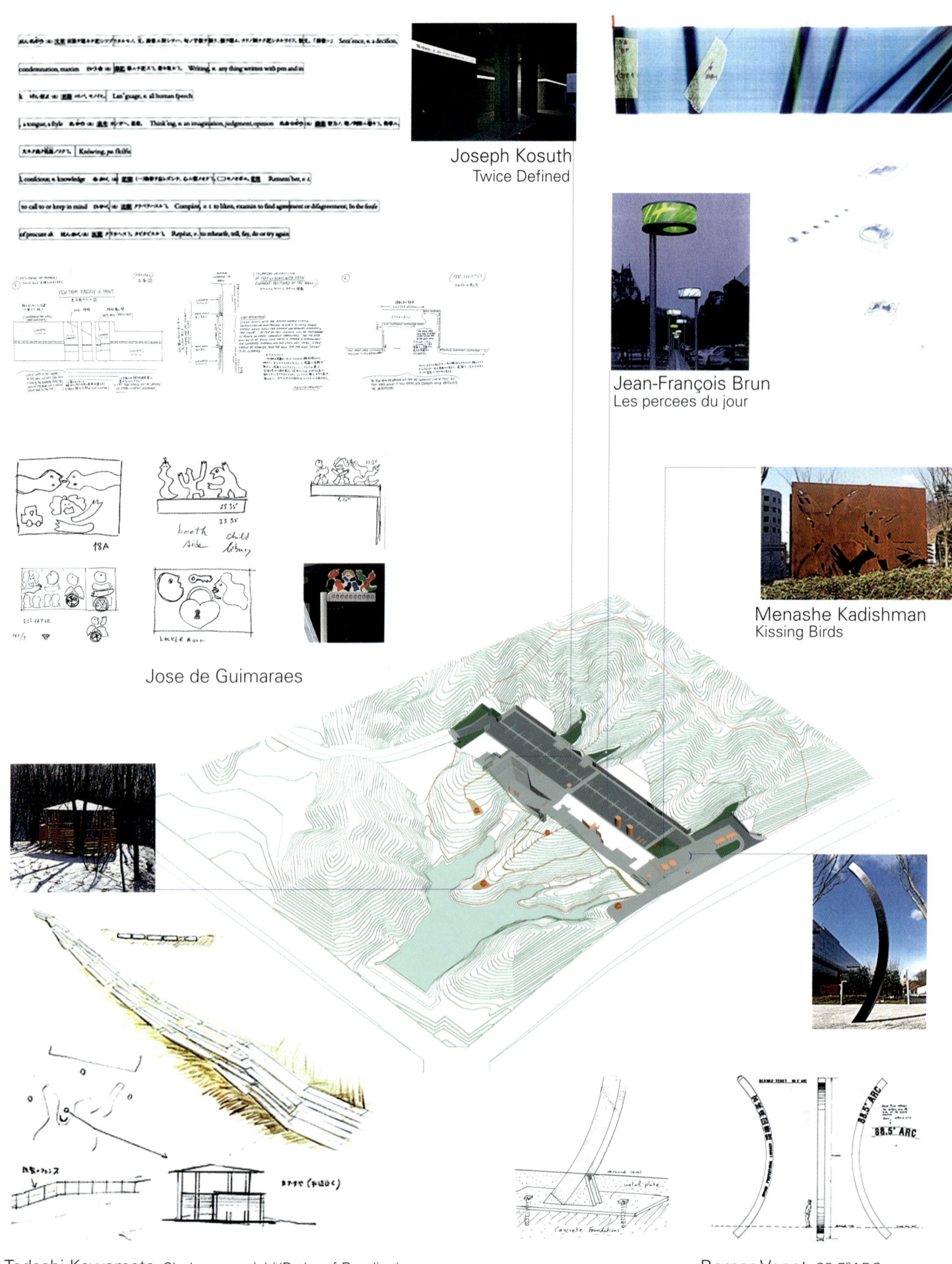

Jean-François Brun
Les percees du jour

Menashe Kadishman
Kissing Birds

Jose de Guimaraes

Tadashi Kawamata *Shoken-no-michi* (Paths of Reading)

Bernar Venet 88.5°ARC

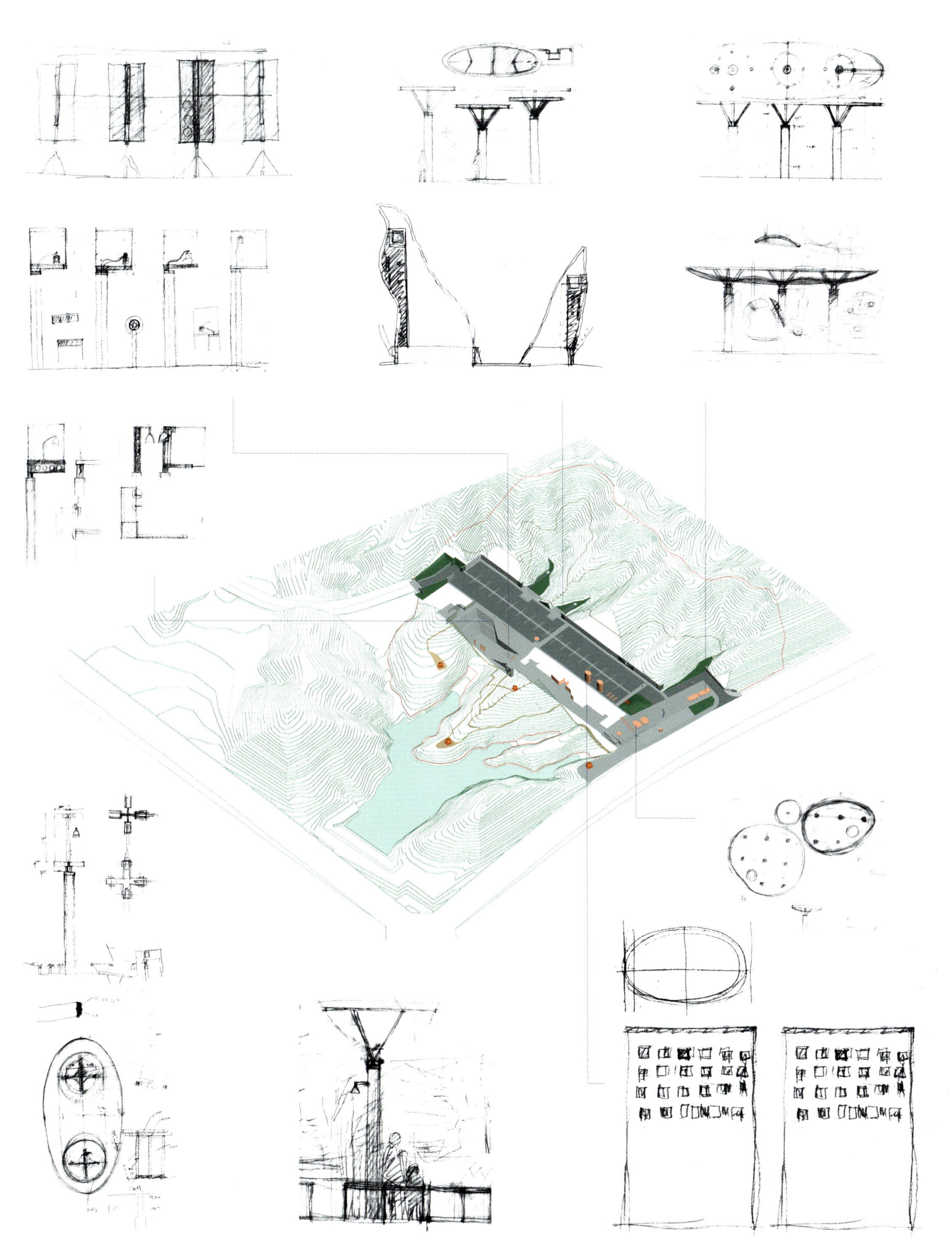

Attractors on the topography in Miyagi Prefectural Library

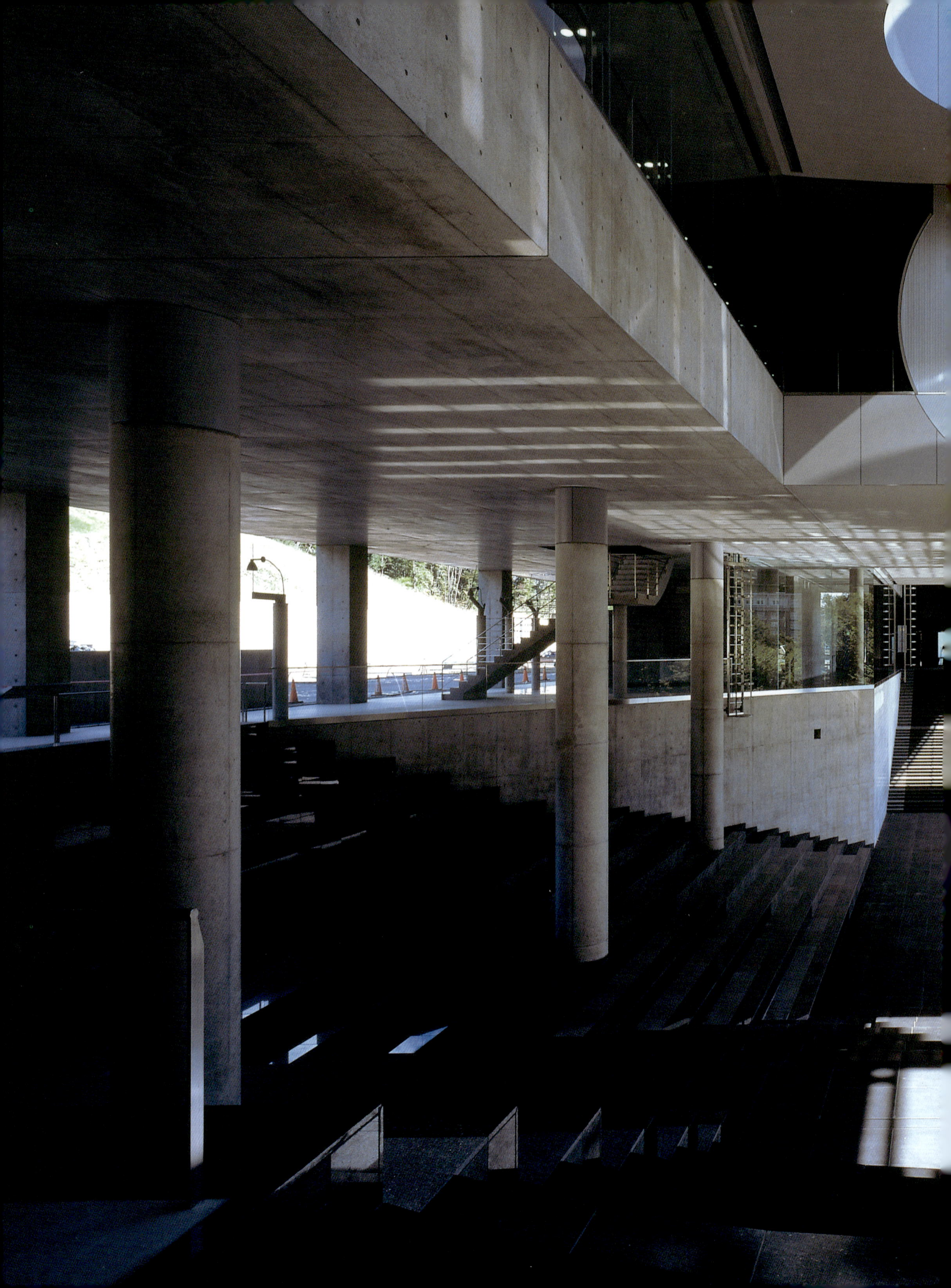

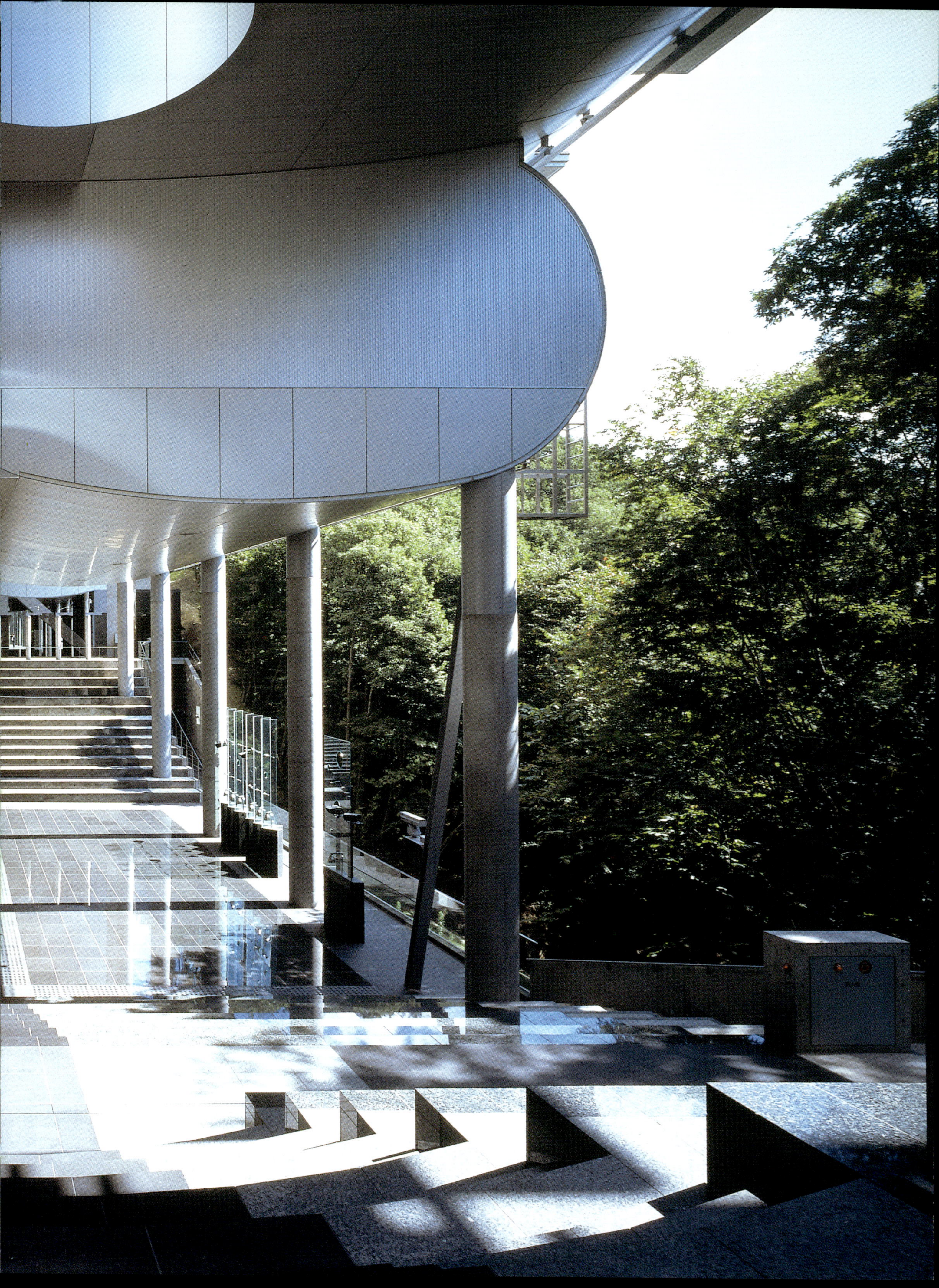

Hiroshi Hara

Previous page: View towards the secondary entrance, seen from the plaza on the first floor

Miyagi Prefectural Library

Interior view of the open-stack reading room on the third floor, shown prior to loading the bookshelves

Hiroshi Hara

Interior view of the open-stack reading room on the third floor, shown after loading the bookshelves

Opposite: View towards the entrance hall, seen from the audio-visual room

Miyagi Prefectural Library

Hiroshi Hara

Kyoto Station Complex, Kyoto, 1997

This building was the subject of a limited international competition in 1990. The station is located in the ancient city of Kyoto, Japan's former capital, which is surrounded by mountains. There were those who argued against the construction of the proposed large-scale building. My position on this as an architect was that in this world, some things are reversible and others irreversible. In the case of architecture, except in those instances in which the natural environment clearly suffers a mortal blow, things can in time be restored and the effects of architecture can generally be reversed. From the perspective of the environment, what we should be doing is to reconstruct areas that have already been built up in cities, not to expand the domain of human habitation.

Whether or not Kyoto Station represents the correct way of reconstructing the city is open to question. However, the area around the station had already been urbanised and was regarded as a future centre of urban activities even before this project got underway. The sheer size of the building is certainly questionable. However, an architect has the responsibility to deal with the task he is given as skilfully as possible.

From a city-planning point of view, I took the following points into consideration:

1. The main blocks of the building should conform to the 1,200-year-old urban grid of Heiankyo.
2. A deck should be built over the railway tracks in the future to prevent the division of the city.
3. The building should be articulated horizontally.
4. The height of the building should be kept as low as possible.
5. The blocks of the building should be arranged and the form of the facade should be designed so that shadows are not cast on the open space in front of the station.
6. A glass-roofed multi-level plaza should be created inside the building since the open space in front of the station will be used mostly by vehicles.
7. The roof of the building should be used as a terrace for the multi-level plaza, which will be made as public as possible.
8. Part of the multi-level plaza should be set aside for civic activities.
9. The upper parts of the building should be finished as far as possible in glass so as to reflect the sky and make the building seem lighter in weight.

I myself am not entirely convinced that all these decisions were correct. With respect to the building as a whole, the intention was to establish a heterogeneous order in which spaces of different quality exist simultaneously. Such an order is the polar opposite of a simple geometric order.

View towards the terrace plaza and the large stairway facing the open space of the atrium

Bird's-eye view from the southwest

Opposite: Perspective of the concourse seen from the glass roof

Kyoto Station Complex

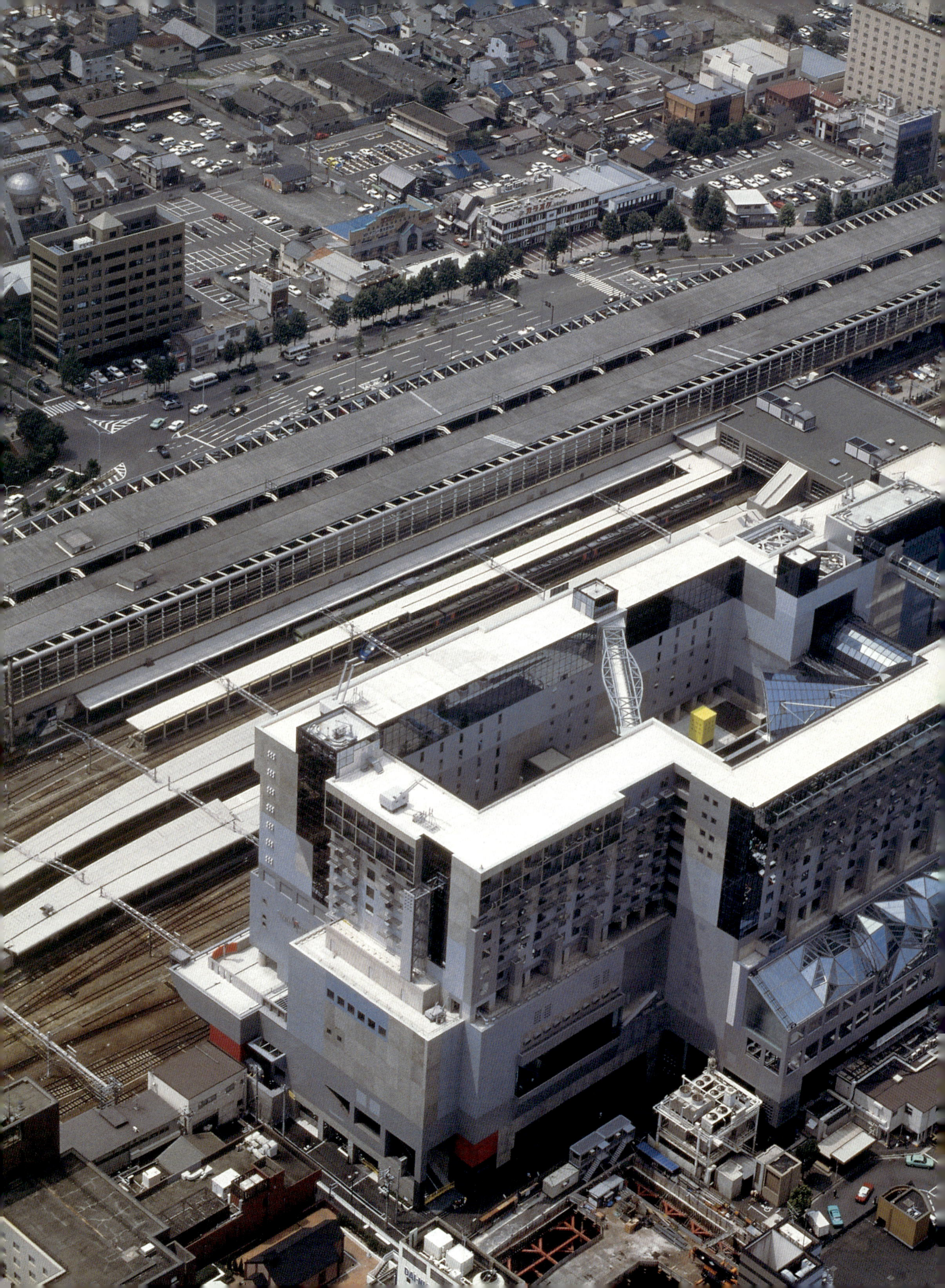

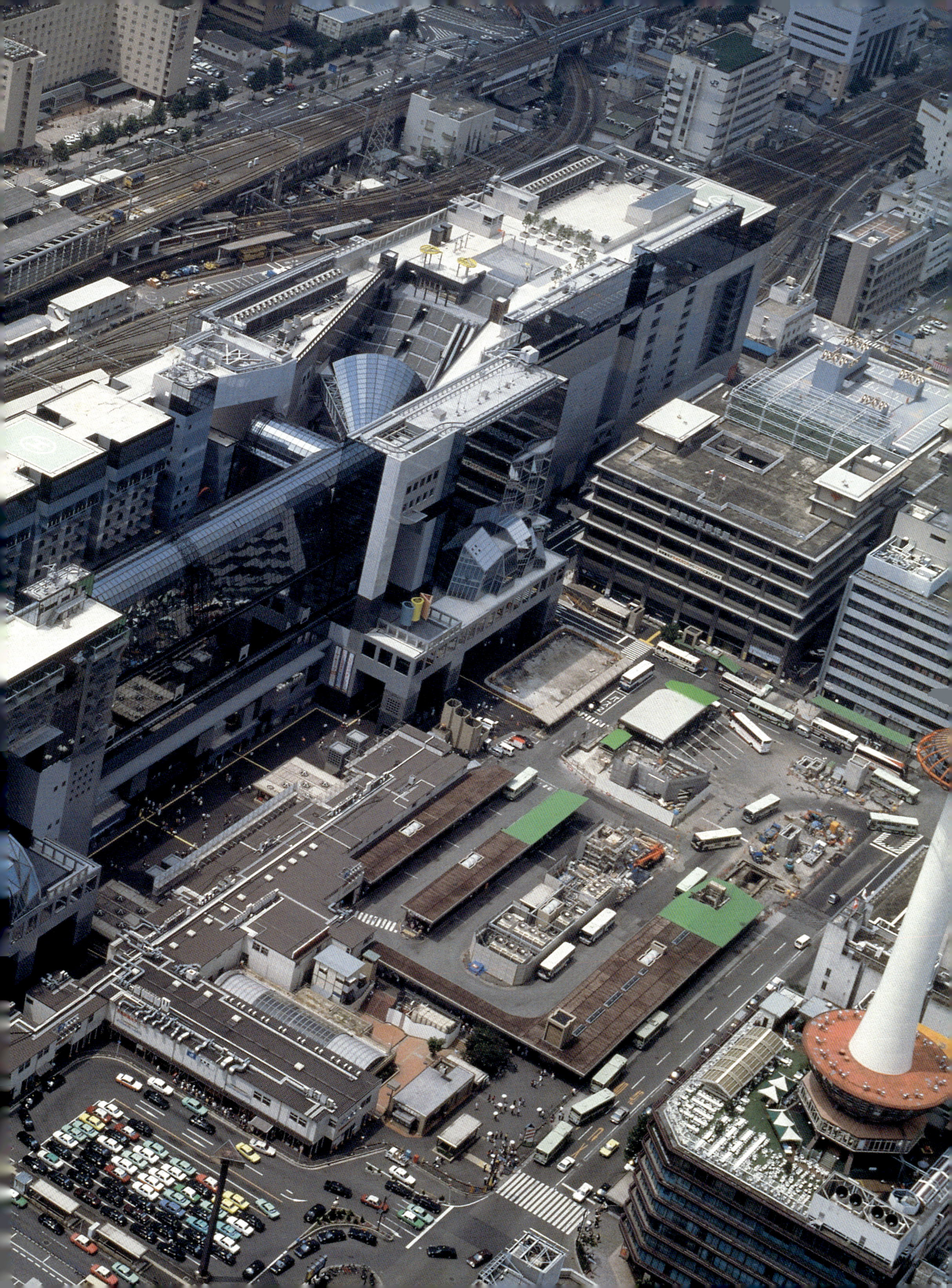

Previous page: Bird's-eye view from the northeast

The north facade

Hiroshi Hara

Kyoto Station Complex

View towards the huge open atrium, an exterior space with a glass roof, seen from the east side of the concourse

Hiroshi Hara

Kyoto Station Complex

Eastern sky seen from the hotel terrace above the concourse. The canopy hangs over the escalator to the Skyway and the mid-air bridge connects hotel corridors

Hiroshi Hara

Kyoto Station Complex

Viewing from the South Square, escalator to pedestrian walkway and Lichtenstein's art advertisement

Opposite: East Square seen from the hotel rooftop

Roy Lichtenstein

Robert Longo

Joseph Kosuth

Tadanori Yokoo

Jose de Guimaraes

Piotr Kowarski

Kokyo Hatanaka

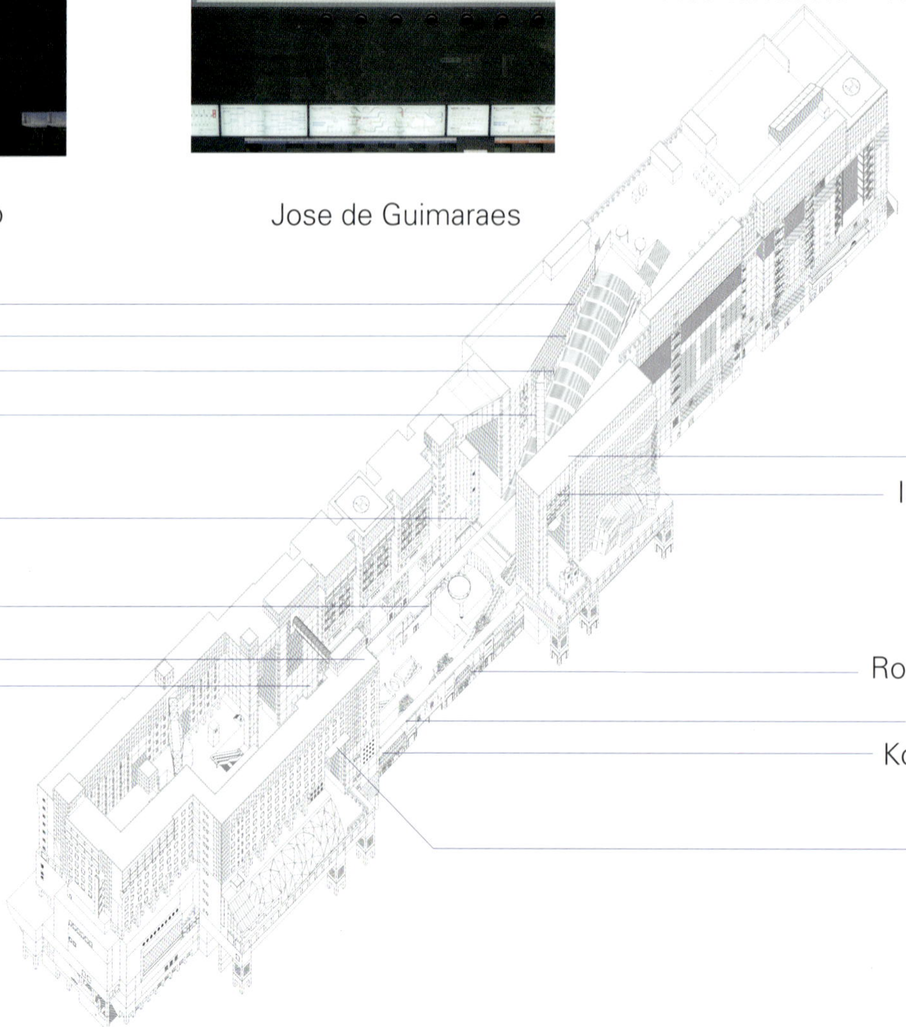

Kojo Tanaka
Miyuki Ueda
Seipu Togawa
Piotr Kowarski

Thomas Shannon

Tadanori Yokoo

Jose de Guimaraes
Joseph Kosuth

Jun Tsuzuki
Isao Yoshimura

Roy Lichtenstein
Robert Longo
Kokyo Hatanaka

Koji Kinutani

Thomas Shannon  | Koji Kinutani  | Isao Yoshimura  | Seipu Togawa  | Jun Tsuzuki  | Kojo Tanaka  | Miyuki Ueda

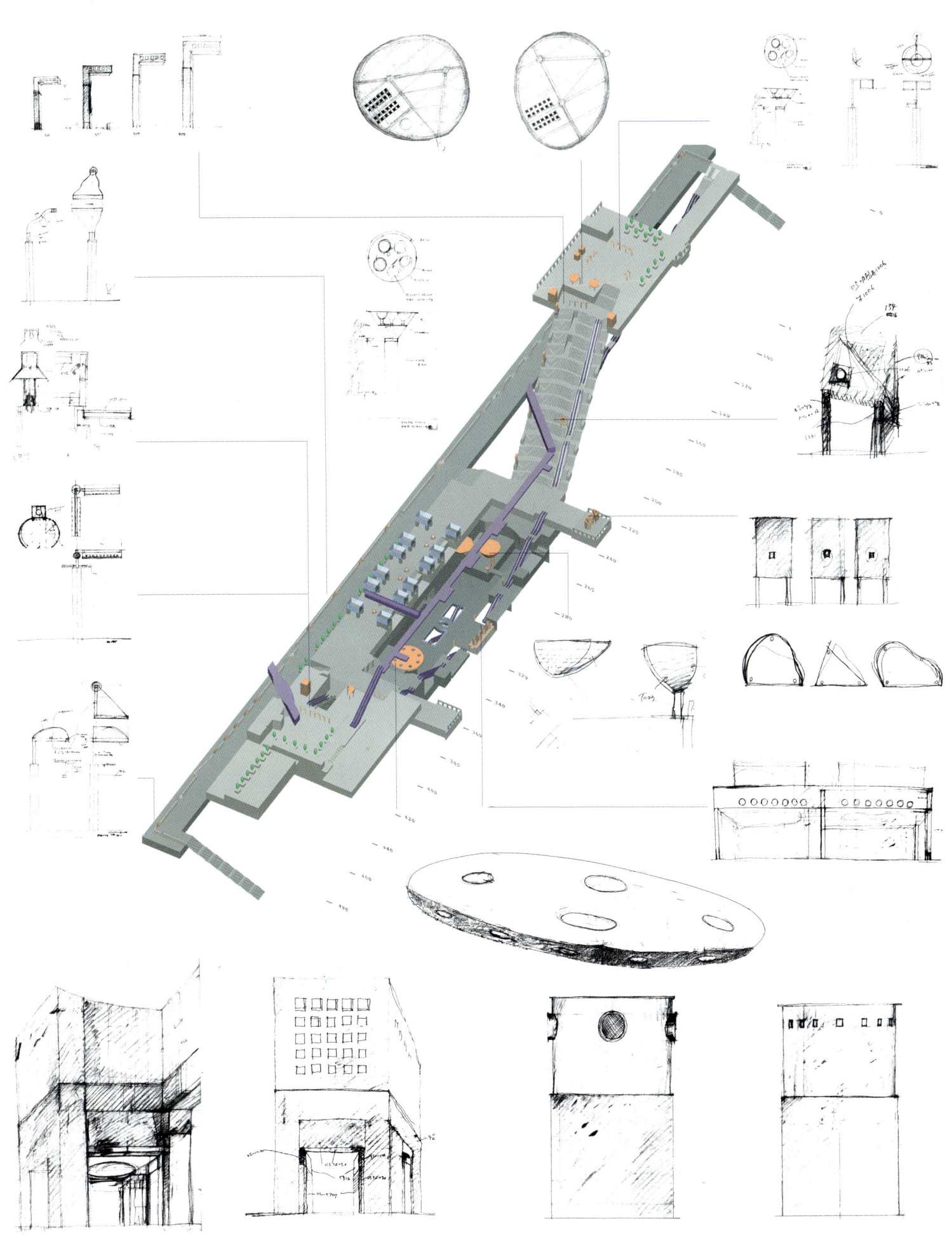

Attractors on the topography in Kyoto Station

Hiroshi Hara

The lobby for the hotel banquet rooms and convention halls

The illumination towers at the entrance of HOTEL GRANVIA KYOTO and Theatre 1200

Kyoto Station Complex

Grand Theatre, the main hall of Theatre 1200. Its ceiling illuminates the flying seats on a diagonal bridge

6. Traversing / Bridging Movement
by Hiroshi Hara

Space appears only when someone experiences it. The phenomenological philosophers were right to say, though there were many aspects yet to be discussed, that space and experience should be considered as one and the same. I would like to call this simultaneous understanding 'traversing'. Traversing is thus to cross both physical and semiotic fields. If all experience involves spatial traversing, then it could be said that there is nothing special about architectural methods or concepts. However, it is still possible to design a building with traversing in mind.

It is almost impossible to make a systematic analysis of human movement in architectural and city planning. (Although it is interesting that in designing a building one can make complicated passages that still function well.) Simply put, in terms of architectural design, people move along a predetermined route. I call this movement 'predictable traversing'. Streets and corridors are designed with this traversing in mind. The second type of movement is the random movement observed in plazas and rooms. This is called 'unpredictable traversing'. It has almost the same meaning as floating. The first represents the necessary mode, while the second represents the possible mode.

However, architects often insert alcoves in streets and windows in corridors in order to change predictable traversing into unpredictable traversing. The Reflection Houses were designed with strong symmetry around a central 'street'. The movement of people in this street was predictable traversing, but I tried to give it the character of unpredictable traversing.

In Kyoto Station (1997), a 25-metre-wide passage forms a V-valley in the building. Human movement is the predictable traversing that goes up and down this passage. At the same time, I located various small plazas along this main passage, as well as paths that connect them in order for people to choose their own way, ie, to change predictable traversing into unpredictable traversing. In the discrete semiotic field, I intended to induce random movement.

In the Reflection Houses, I placed a glass roof or a row of skylights above the passage of predictable traversing, in order to admit natural light. With the sky and sunshine, the indoor corridor would change into malls and streets. This approach emphasises passages rather than rooms. It also emphasises 'traffic' in architectural and urban planning, a concept introduced by Gottfried W. Leibniz (1646-1716), which includes modern-day communication. A spatial structure is given to the field when you open up a relatively closed space and let the sky peek in or the sunlight enter. This is due to the contrast in brightness. If you make this place a traffic junction, then its significance matches the spatial structure. The building becomes legible, even dramatic.

In the Ose Junior Middle School project (1992), the use of traversing was important. The valley of Ose is home to the Nobel prize-winning writer Kenzaburo Oe, who was educated in this very school, but

Diagram of traversing (from 'Grammar of Space')

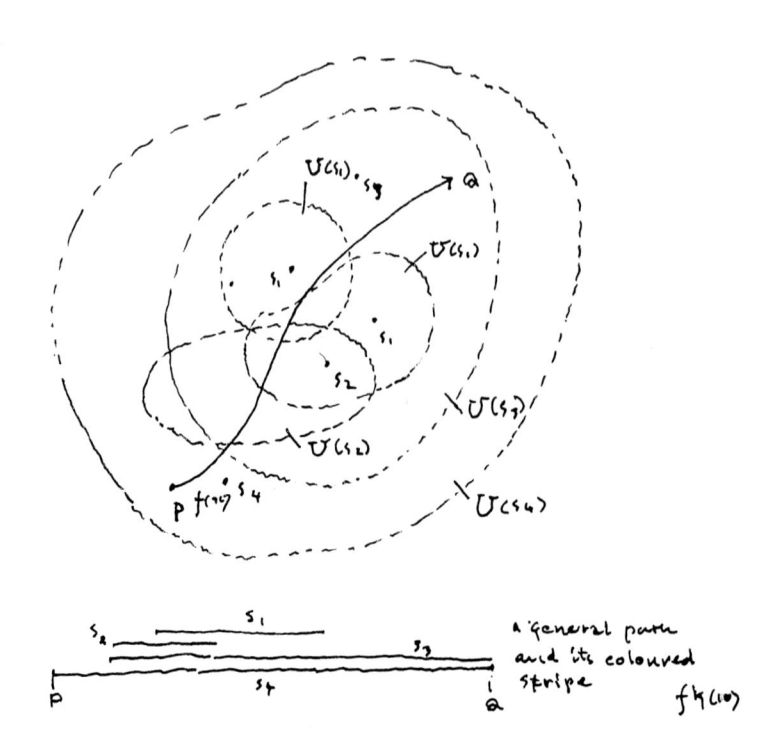

in the old building. Many of his stories take place in the valley of Ose. According to my analysis, in Oe's literature, the valley consists of about thirty 'symbolised' places, among which his stories evolve. Plots vary from story to story but they share the same places, which are mutually discrete and function as attractors; so the characters' movements comprise unpredictable traversing through these places. They mostly correspond to places in the real valley, including the school site. Since Oe is one of my best friends and mentors, my concept was to design the school as an extension to his literature – as a set of symbolised places. Good examples are the cylindrical rooms, which are both studios, one for music, the other for art. The rest of the classrooms are characterised by their roofs. Next, I brought in virtual places that only appear in Oe's stories. One is a 100-metre-long road imported by one of Oe's characters from Mexico's Teotihuacàn. I used this as the passage along the patio and playground.

The passage facing the patio starts among Ose's traditional tiled roofs and ends at the cylindrical music room. It is a passage of predictable traversing. The traversing sequentially bridges Japanese domestic architecture and the international. Although it shares the same length, it has no other connection to Teotihuacàn. It is made of the old building's roof tiles and stones from the river bank.

The building, situated on a slope, consists of three different levels, connected by two bridges. Students can unpredictably traverse among various 'symbolised places', but can locate their positions by going out to the bridges.

Kenzaburo Oe wrote two long stories after the completion of the school, which is mentioned in both. This is a fictional traversing between literature and architecture.

The Miyagi Prefectural Library (1997) is a bridge crossing a valley. The main library resembles a mall, as if a street were located in the building. Another passage traces the topography on the ground, so the bridge induces two layers of passages. In the Umeda Sky Building (1993), a symbolic bridge connects two skyscrapers. Bridges appear whenever architecture is considered as urban topography. The more complicated the topology, the greater the number of bridges. In Kyoto Station, in order to make various networks of passages, many bridges were installed.

The Motomachi High School, which has a long history, is situated in the middle of Hiroshima city on a limited site. Since classes had to be continued in the old building during construction and the floor area needed to be twice as large as the original building, I placed art studios above the new gymnasium, which I considered as a small plateau, an urban topography. The other leg of the L-shaped building, occupied by classrooms, had to be densely packed. I therefore sandwiched a valley using two buildings and let the light in from the top. The students first ascend to the fourth, plateau level via an escalator

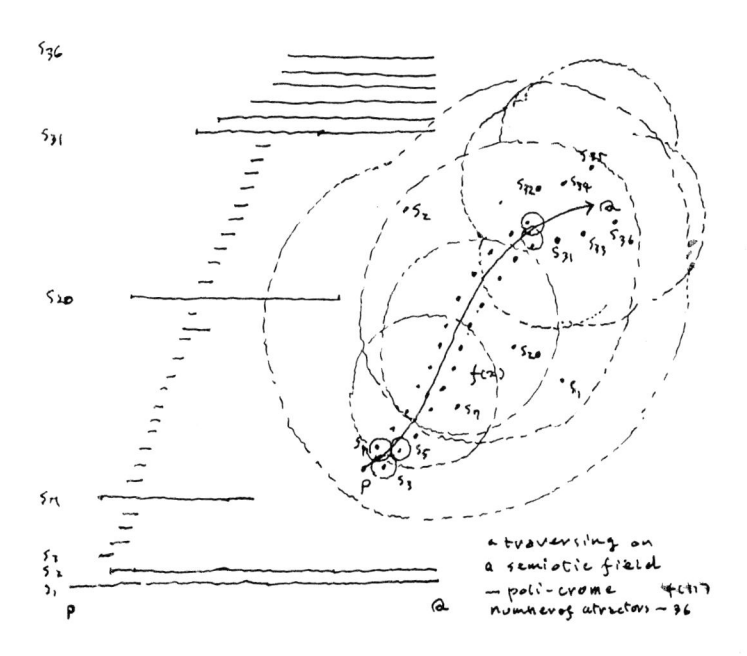

Diagram of traversing (from 'Grammar of Space')

then disperse to the classrooms. The valley is crossed by several bridges, where students can survey the whole building from the ground to the fourth level.

The high school has sixty classrooms, which are of three types: normal, special and studio. Several variations within each type make every classroom unique. Starting from these classrooms, much traversing takes place from playground to gymnasium to lecture hall and other destinations.

The passages and corridors of the classroom annex are based on predictable traversing. However, escalators and bridges form various loops so that one can choose from many routes. The movement of children is therefore complex, almost unpredictable traversing. Since all the classrooms are located on the fourth level, it is always busy.

This programme of traversing maximises student interaction. For high-school students, making friends and socialising is the most important aspect of their schooldays. With 1,000 students in each grade, every student has a chance to make 5,000 friends. One of the most crucial targets of traversing is to induce this kind of traffic.

Locating the gymnasium and the lecture hall on the same level is also part of the traversing programme. The hall can be used as an extension to the gym, or the gym can serve as a backstage area for performances and events. Here, the traffic is between the two areas in terms of their usage.

A bridge can connect two areas while maintaining their independence. During predictable traversing, when one comes to a bridge the view expands, thus making it easy to navigate. Moreover, bridges imbue the building with a floating atmosphere. This is why I often include bridges and naturally lit malls in my work, especially in a dense environment such as the school in Hiroshima.

The Komaba II campus (still uncompleted) of the University of Tokyo is even more dense than the school in Hiroshima. Here, I devised a masterplan whereby the old buildings would gradually be taken over by new ones. A high density of laboratory facilities will encircle a plaza, following a basic system of urban facades and pilotis. The labs are accessed through patios and atriums. The original intention was to cover the pedestrian areas with glass roofs, but the programme would not allow it.

The basic concept of this facility was to create a university that was 'open' to the public, but since the hazardous high-tech labs do not allow public access, it is 'clopen' – open and closed at the same time. The plaza and arcades are public, unpredictable traversing areas, but the pilotis act as thresholds. The space under the glass roof of the Institute of Industrial Science was intended to be visually continuous, but was articulated into four parts to meet safety requirements. These articulated malls are defined by the facades of the individual labs, which can be seen from the piloti.

Takagi Clinic (1999) is located far from urban areas and from the traffic point of view is like a bus stop in the countryside. Since it specialises in obstetrics and internal medicine, it tends to cater for babies and elderly people on short-term stay. The roof was designed taking into account the snowy climate. Through a road surrounded by ample greenery one enters the lobby and encounters a surprising view caused by tricks of perspective. Local people use the building in their everyday traversing as a venue for small concerts and exhibitions.

Ose Middle School elevation

Traversing / Bridging Movement

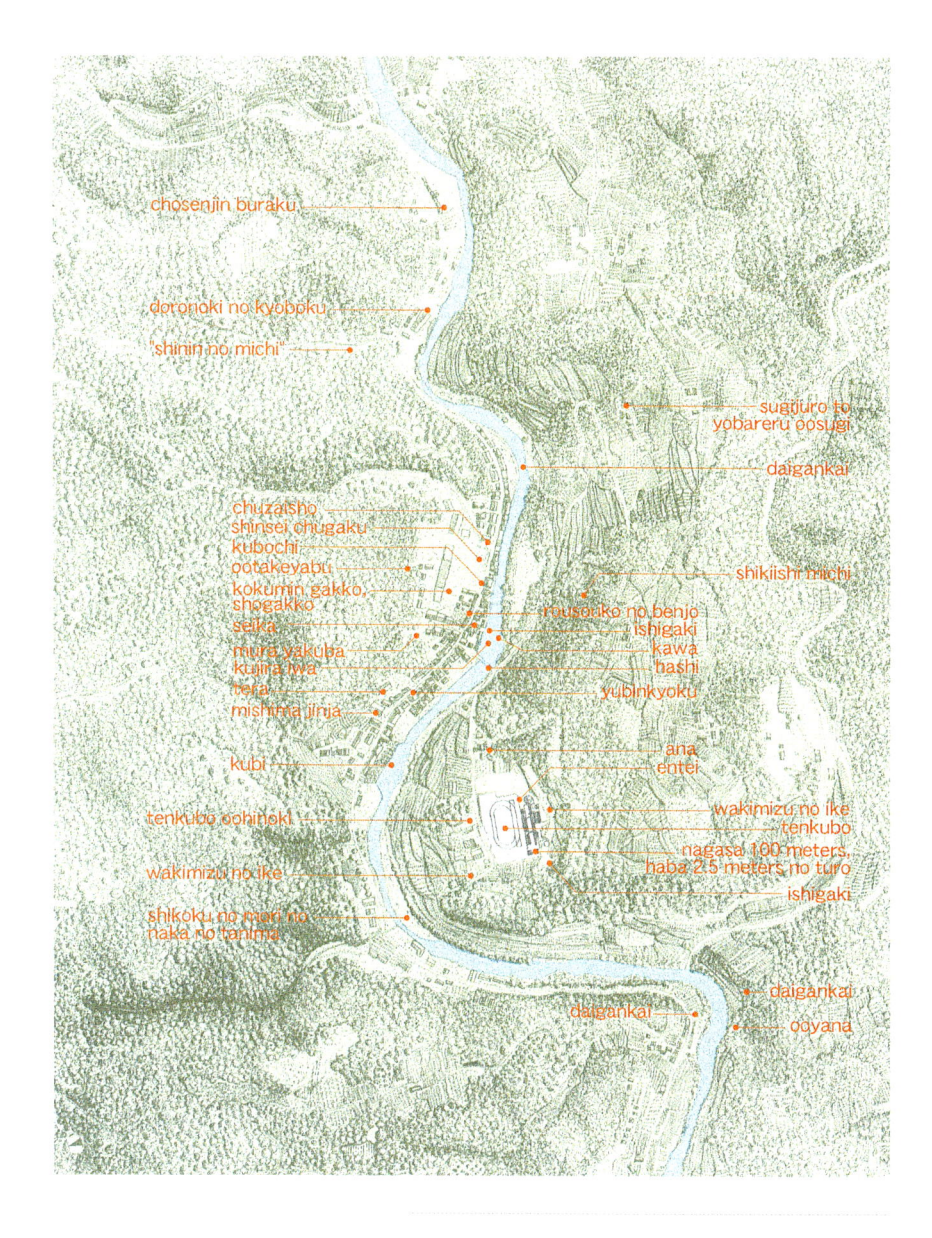

31 semiotic places in the valley of Ose in Three Novels *by Kenzaburo Oe*

Ose Middle School, Uchiko, Ehime Prefecture, 1992

The writer Kenzaburo Oe, whom I admire enormously, has described Ose, in which he grew up, as 'a valley in the forests of Shikoku'. The community is small, as is the valley. The middle school contains only about 100 students and was therefore designed not just as a school, but as a small cultural and sports centre for the district. That is, rooms such as the music room, art studio, library, audiovisual room, home-science room and science room have been arranged around the lobby, which can function as a gallery, so that they can easily be used for various events held by the village. Locating the building on the top of a slope made the athletic field more spacious. Having grown up in a valley, I could well understand the wish of the people of the village 'to play baseball on a spacious field for once', as they put it.

There is no way of predicting the significance that cultural activities in Japan in the latter half of the twentieth century may have for the future, but for many years Oe, the composers Toru Takemitsu and Toshi Ichiyanagi, the architect Arata Isozaki, the theatre director Tadashi Suzuki and myself have met for discussions that transcend the concerns of any one genre. We all belong to the same generation, still children at the end of the Second World War, and growing up in the free democracy of the postwar era. Unlike the previous generation, which included the filmmaker Akira Kurosawa and the architect Kenzo Tange, we were able to study intellectual activities in the modern West and artistic expression all over the world. This intellectual environment was made possible by a large group of able translators. As far as artistic activities are concerned, our generation might be characterised as the 'method generation'. No artistic activity is of course possible without intuition and sensibility. For example, Takemitsu's music reveals an extremely refined sensibility. Yet a distinctive aesthetic method underlies his major works. This suggests the great influence of Russian Formalism, Surrealism, Marcel Duchamp and John Cage, which was shared by us all. Oe's body of work is also based on a grand concept.

That concept was nurtured by the small valley of Ose. Those interested in Oe's work will no doubt continue to visit the area far into the future. *HH*

View towards the cylindrical music room, seen from the courtyard

Opposite: Perspective of towards the courtyard

Overleaf: Bird's-eye view from the north

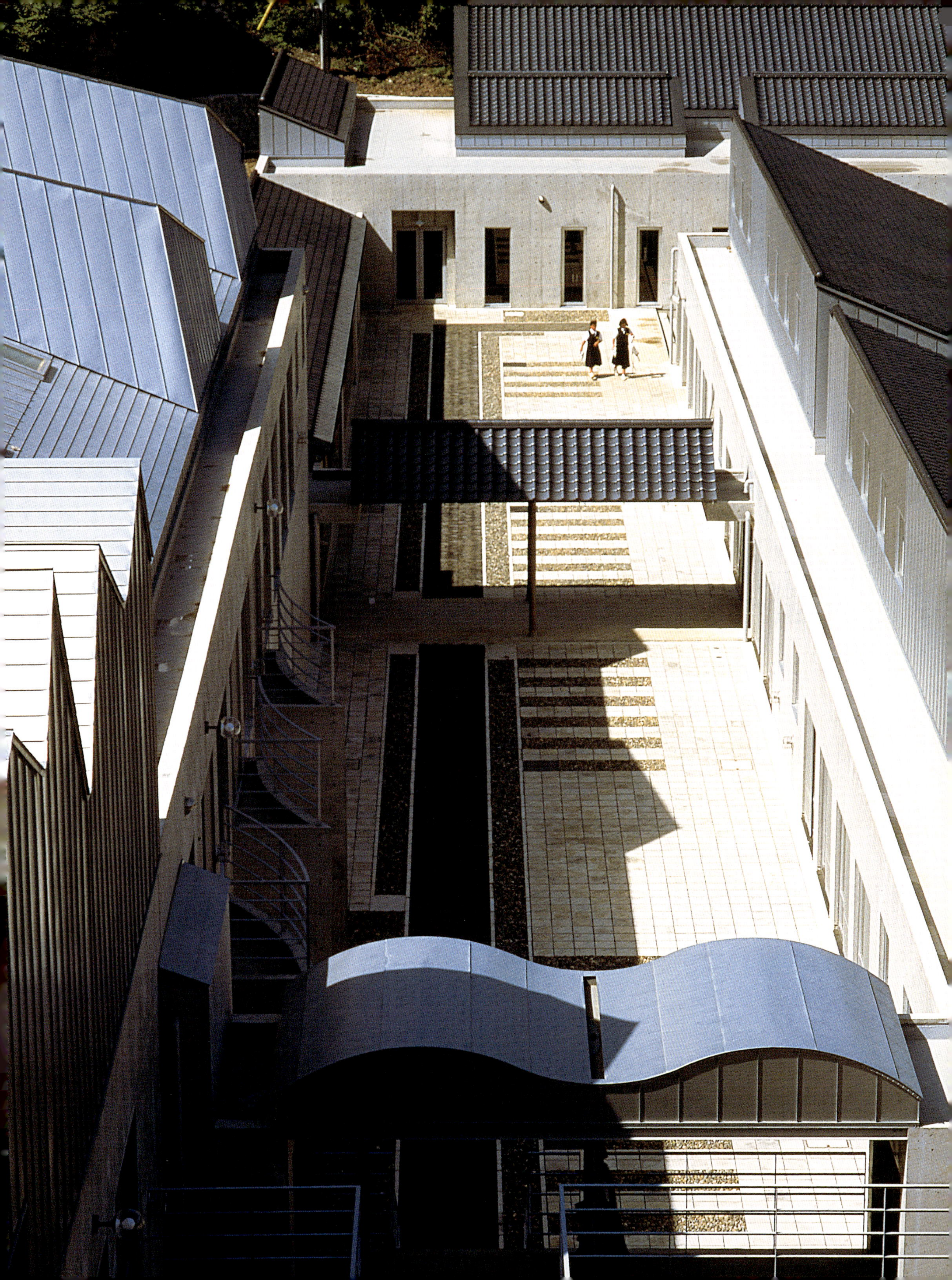

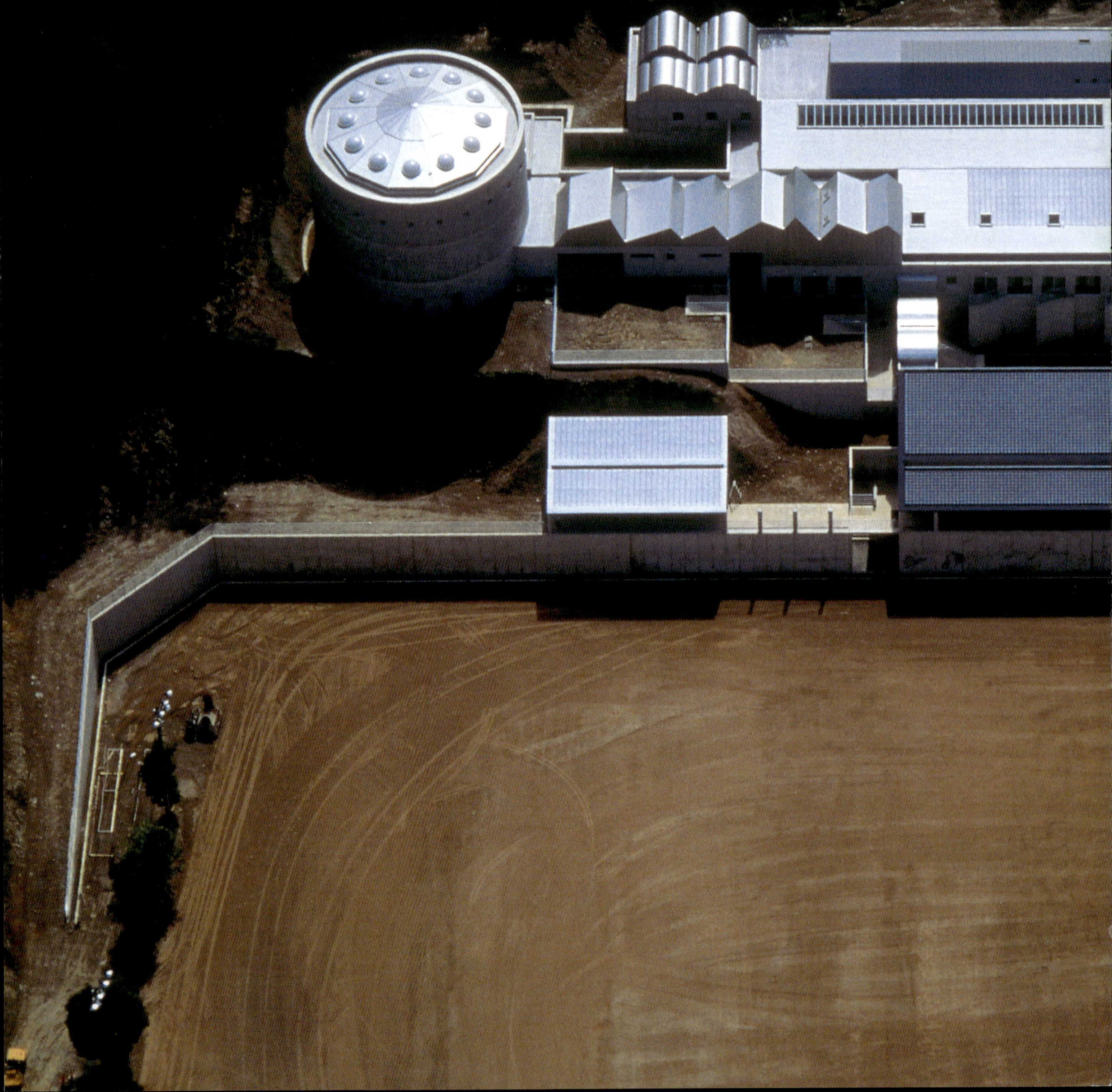

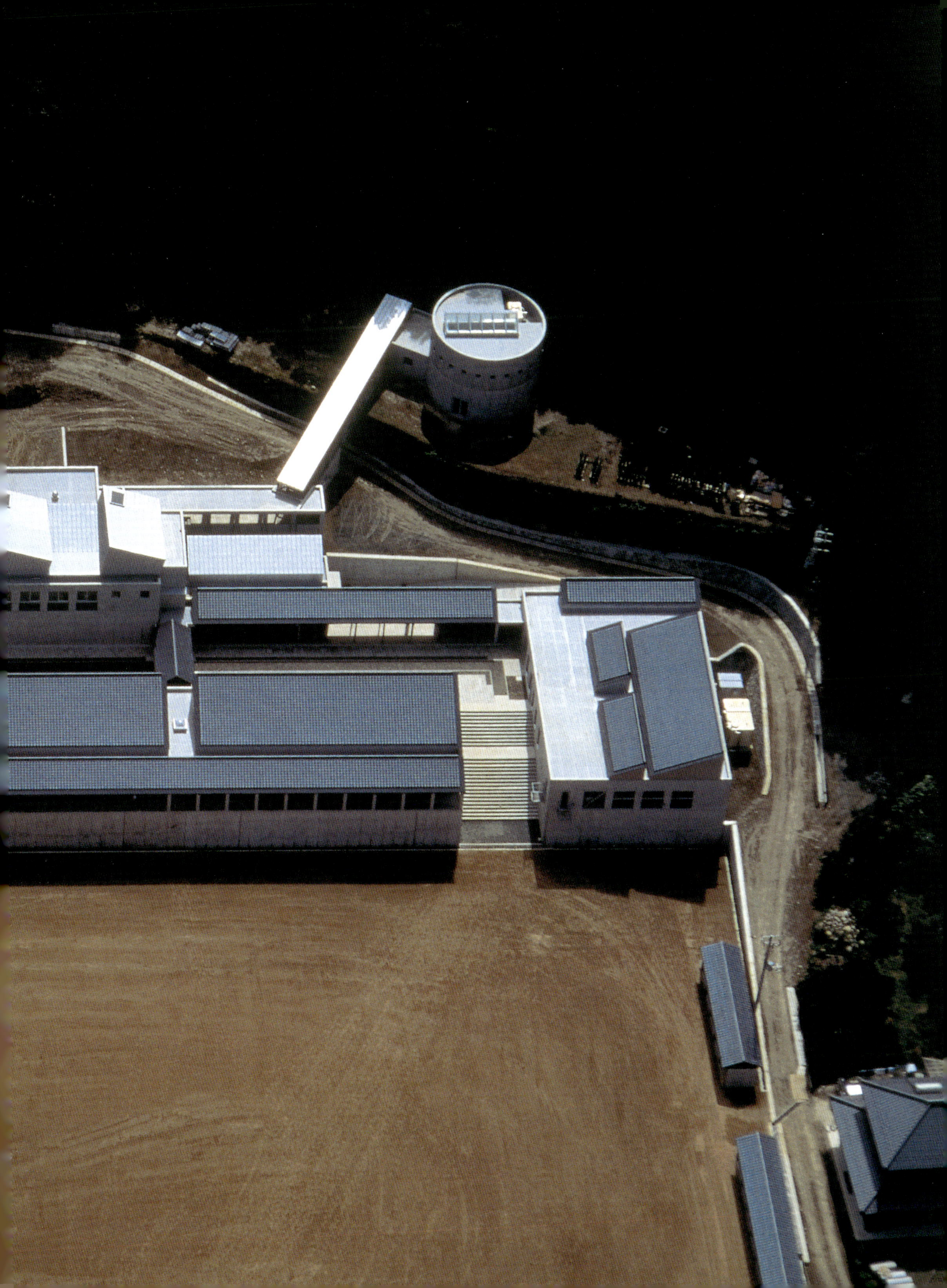

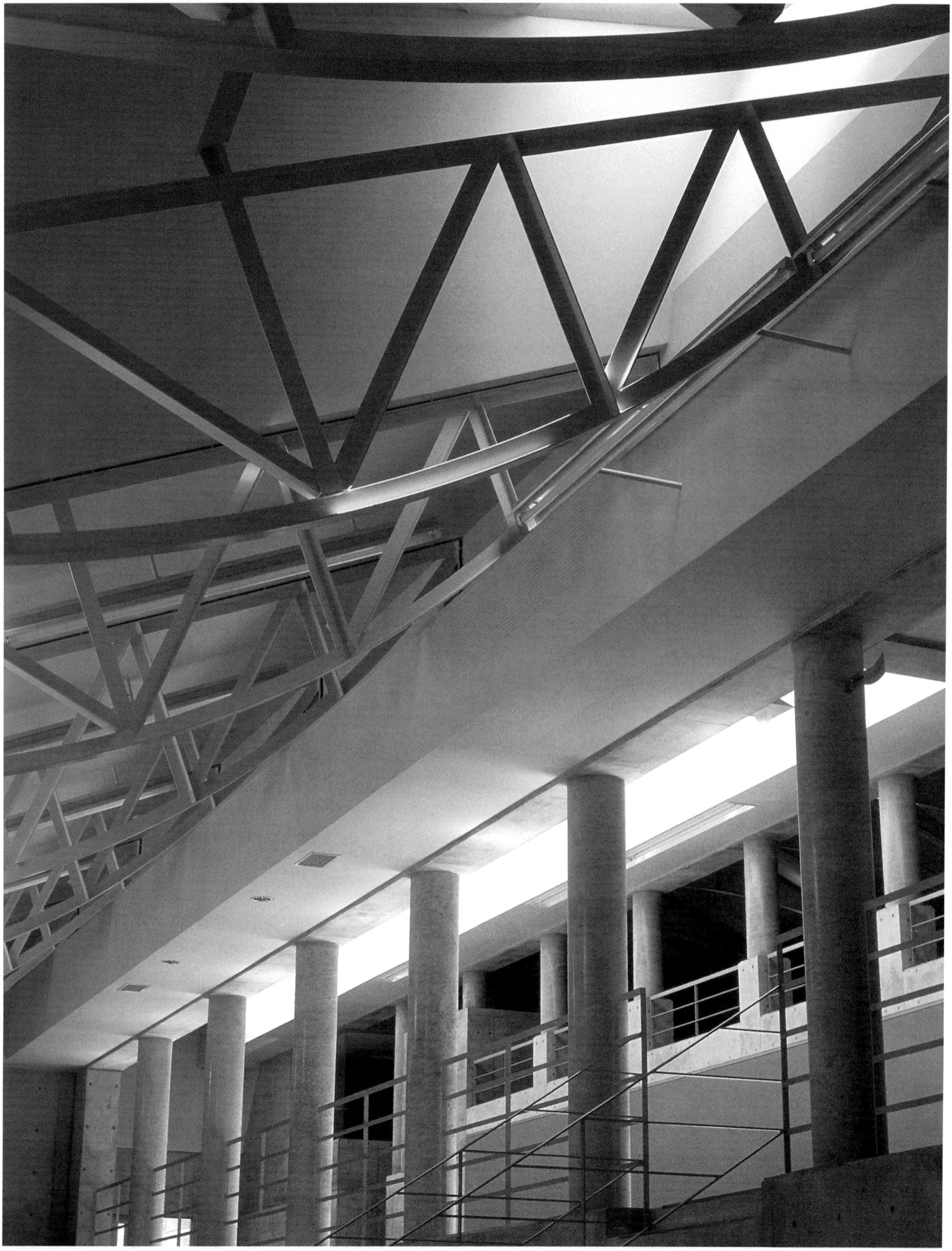

Ose Middle School

Exterior view from the west

Opposite: View towards the multipurpose hall, seen from the science room

Overleaf: Overall view from the playground

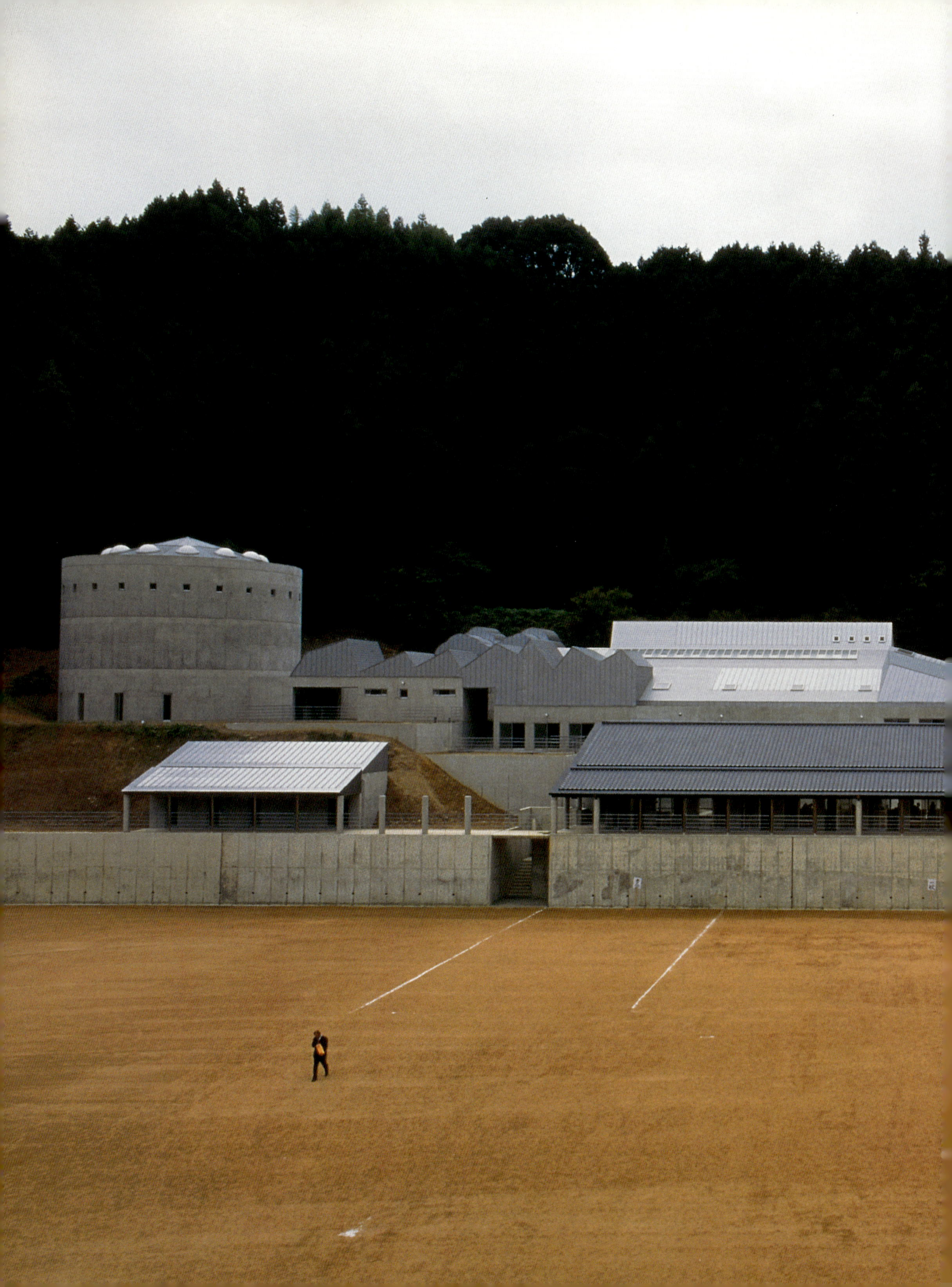

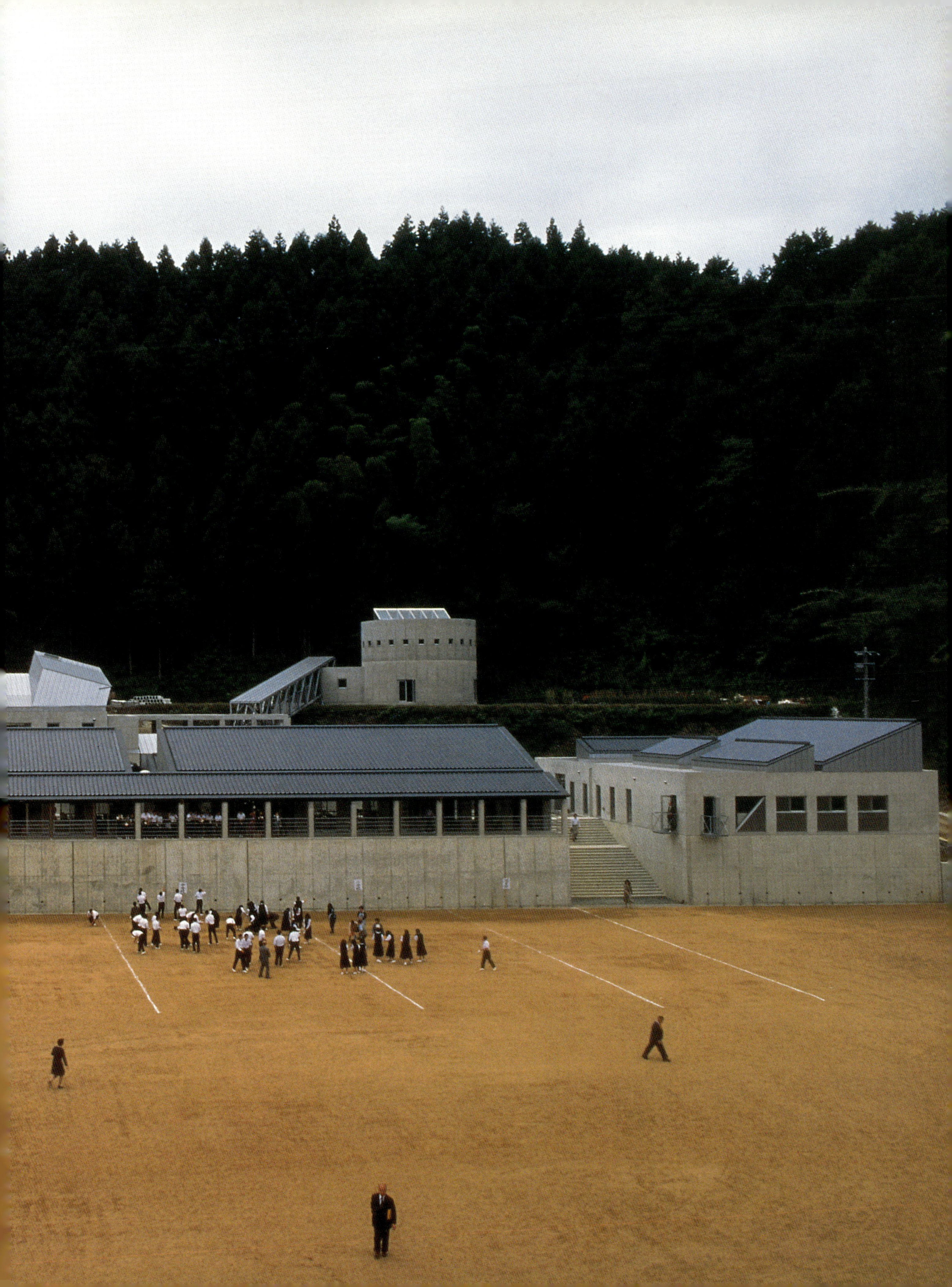

Hiroshi Hara

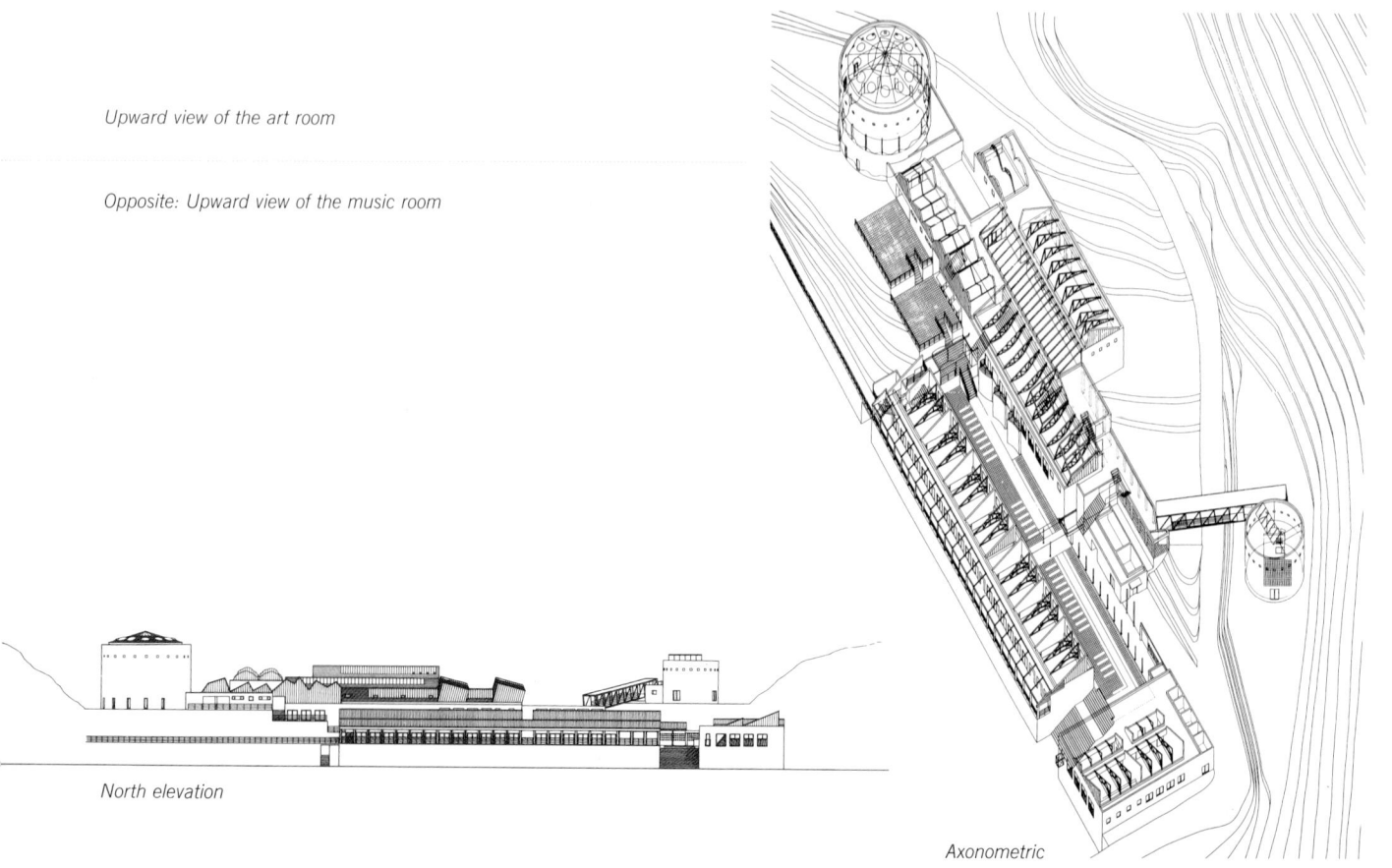

Upward view of the art room

Opposite: Upward view of the music room

North elevation

Axonometric

214

Ose Middle School

Motomachi High School, Hiroshima, 2000

Hiroshima city was wiped out by atomic bombing during the Second World War. Over half a century later, it has developed into a modern city. Under the slogan 'Peace and Create, 2045', it is carrying out a programme to create a number of excellent buildings by 2045. This building belongs to that programme.

The site is in front of the reconstructed Hiroshima Castle and the surviving castle moat. The school was made continuous with the park on which the castle stands – the fact that this is a municipal school made this link possible – and the pilotis of the school is intended to serve as a shelter for the park. Since the school facilities had to be crowded together, topography was incorporated into the design to make the classrooms more distinctive.

In the years since the Second World War, school buildings have undergone greater changes in form than any other building type in Japan. (Apartment buildings have changed the least, a contrast that reveals a great deal about the character of Japanese society.) One reason for this was that Japan underwent a rapid process of modernisation, and people had a better understanding of the importance of education. However, the diversity of school architecture is mainly attributable to the fact that responsibility for constructing school buildings for all grades up to the senior high-school level was shifted from the central government to local governments. Uniformity in school buildings is a thing of the past. The introduction from Great Britain of the 'open school' form of building also contributed to a reassessment of the mechanical arrangement of classrooms.

However, individual classrooms have generally changed very little. My main concern in this school was the design of each classroom. Air conditioning was introduced because there are many classrooms in close proximity. The air conditioning and mechanical service ducts and lighting equipment were left exposed to create a higher ceiling in the studios for the course in creative expression.

From the beginning of my career as an architect I have made it a point to design the whole from the parts. In this case, what is ordinarily the whole is nothing more than a part. The school is a collection of diverse parts and the whole can only be perceived by an observer traversing the facility. Such a conception of the whole is also evident in the organisation of indigenous communities. *HH*

Interior of the exhibition gallery on the fourth floor of the west wing

Motomachi High School

Interior views of the art studios

Hiroshi Hara

Motomachi High School

View of the fourth floor of the south wing

Hiroshi Hara

Above: The south facade

Motomachi High School

Below: The west facade

Hiroshi Hara

Motomachi High School

Interior view of the auditorium

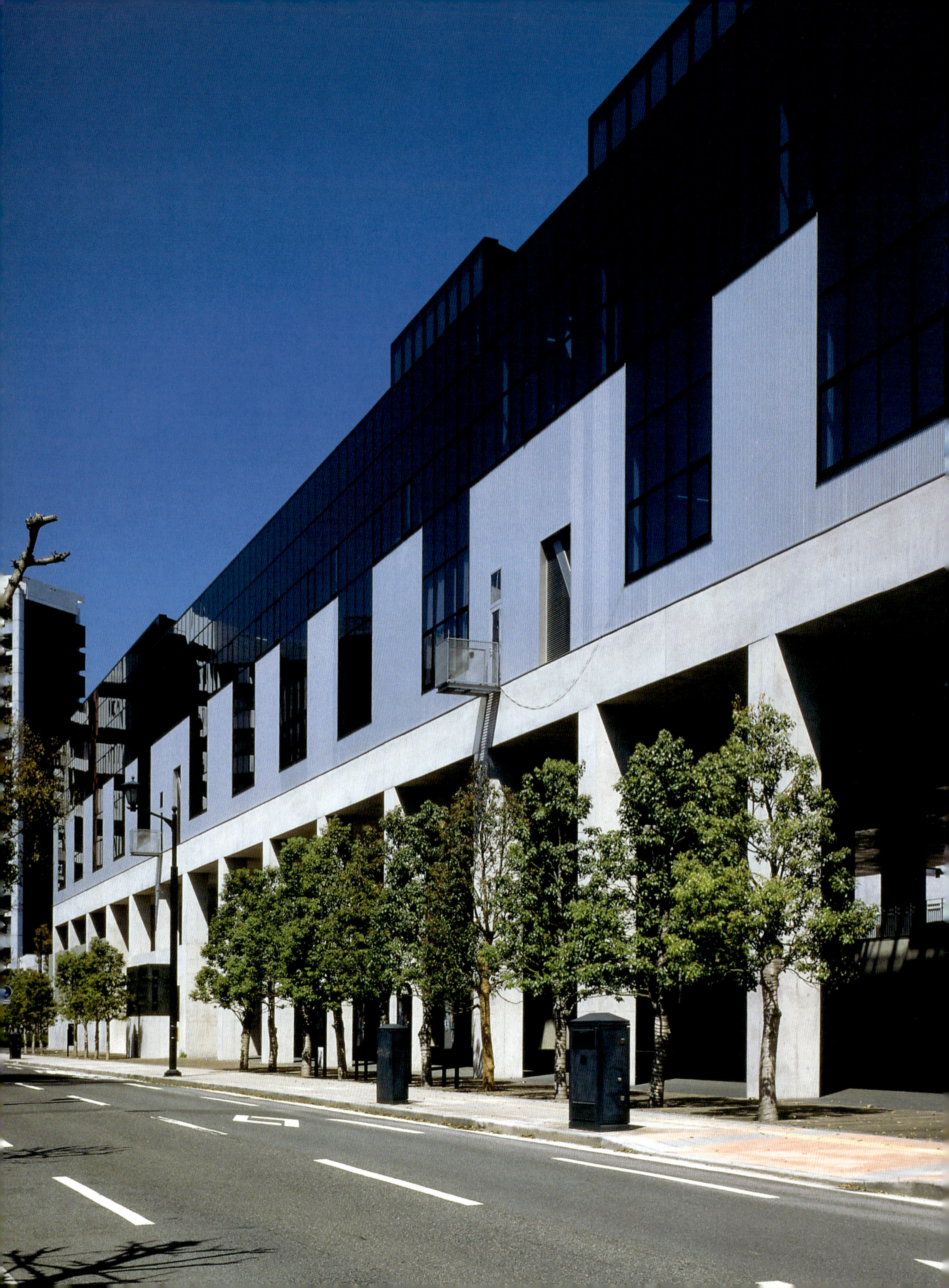

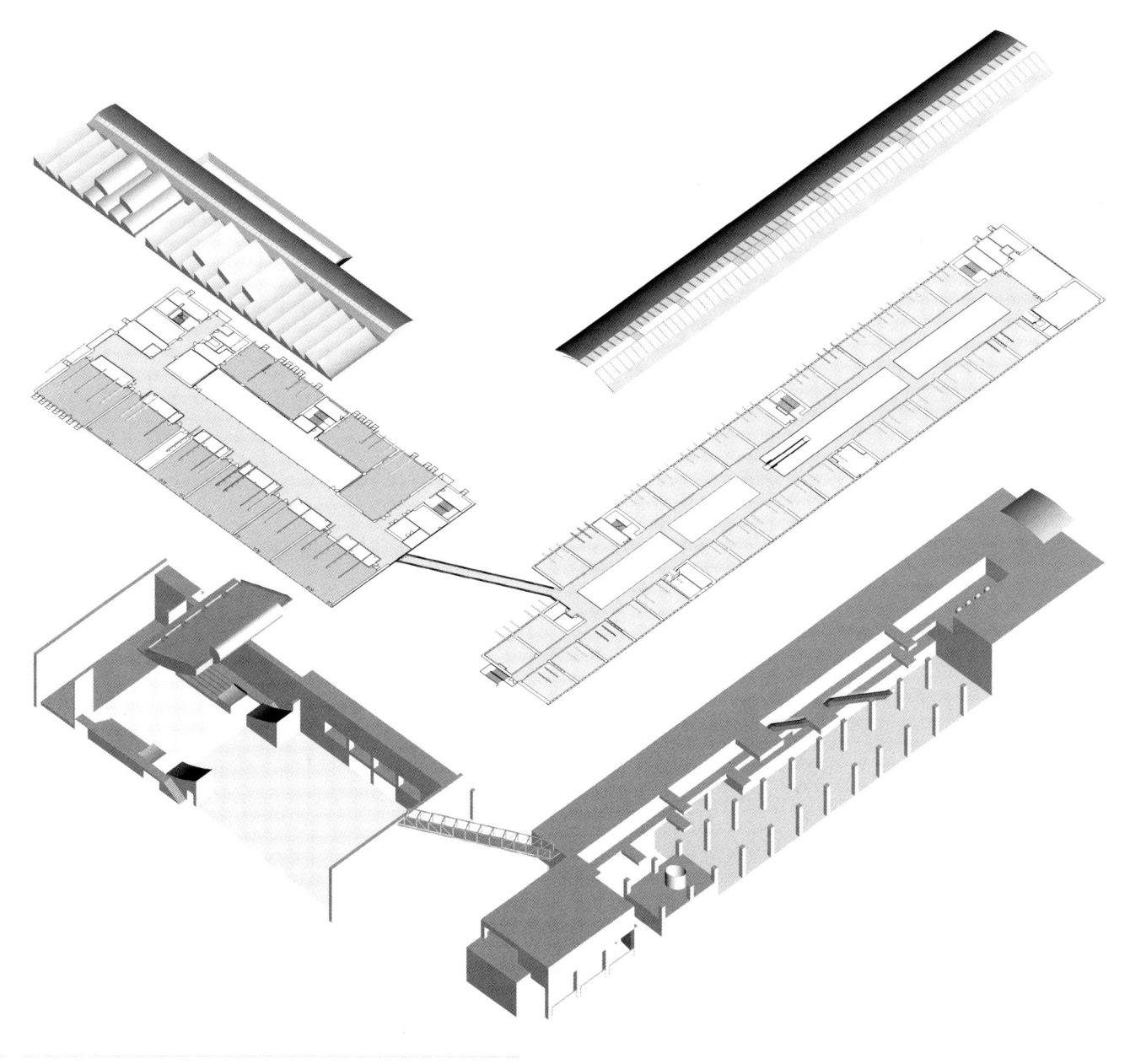

Above: Interior views of the atrium.
Opposite: The south facade

225

Takagi Clinic, Tokamachi, Niigata Prefecture, 1999

Tokamachi is a hilly area on the west side of the country, facing the Sea of Japan. It is in a region subject to heavy snowfall averaging 3.3 metres in winter. What architectural form best deals with snow? There is no easy answer. Basically, there are two approaches: one is to make a structure that can withstand the load of accumulated snow and the other is to create one that can shed it. Both have certain shortcomings. In the case of this building, a clinic with no more than twenty beds, I adopted the latter form: a metal-sheet roof with a pitch of 25 degrees. Even a small amount of snow quickly slides off the roof.

Hospital buildings do not generally have any special character. Under the pressure of demands for rational, functional, easily administered buildings, architects tend to adopt standard forms of hospital architecture. However, I have learned from surveys of indigenous communities that the concept of standard forms is not really applicable to architecture. Standard hospital buildings are those in which even healthy visitors seem to get sick. The designs of such hospital buildings take into consideration only the human body and are themselves machine-like. A human being also has consciousness, which, inseparable from the body, needs to be taken into account. One of the tasks that architects set for themselves in the latter half of the twentieth century was been to create designs for the consciousness while providing shelter for the body. Consciousness emerges when people traverse space. We cannot, of course, design other people's 'consciousness'. However, we can take it into account.

I was under no illusion that the design would have a significant effect on what happens in the building. It is, after all, a small clinic, specialising in obstetrics. I wanted the building to be at least cheerful. The generous space of the lobby and waiting room is a violation of the rules from the standpoint of standard hospitals, where the objective is to make the interior climate as uniform as possible. *Hare* and *ke* are opposing concepts of modality in Japanese tradition. *Hare* refers to the festive and *ke* to the everyday life that people must endure. Sickness is of course a matter of *ke*, but childbirth and recovery are *hare*. The two concepts indicate states of consciousness. How can sickness be cured in a space that is *ke*? How can we celebrate a birth and wish the newborn child a bright future if the event takes place in a space that is *ke*?

The slight twisting of the plan made detailing the framing of the roof somewhat difficult. However, people understand and appreciate that extra effort has been made to create a unique building. HH

View towards the waiting hall, seen from the salon

View towards the waiting hall, seen from the lounge hall in front of the reception area

Takagi Clinic

General view from the northeast

Takagi Clinic

University of Tokyo Komaba II Campus, 2001

As a part of its development policy, the University of Tokyo is rebuilding its Komaba campuses, two adjacent sites located not far from Tokyo's central district. As a member of the university faculty at the time, I was involved in developing a master plan for the Komaba II Campus. Construction is currently taking place on a group of buildings, and more will be built in increments in the future.

The master plan is a programme for structuring the transformation of the campus; that is, for demolishing old buildings and constructing new ones. Buildings to be preserved are concentrated in the area around the approach to the campus. A public area, the University Plaza, will be built in the centre of the site. Giving the public access to the plaza is crucial to efforts being made to achieve an 'open' university.

Several buildings accommodating research laboratories are being constructed at high density around the plaza. Since each building will have a pilotis, the plaza will be surrounded by an arcade. Atriums and courtyards will be connected to the pilotis. Someone traversing the space along the pilotis will see a facade separate from that facing the plaza; that is, the plaza will be enclosed in effect by two layers of facades.

The building for the Institute of Industrial Science (IIS) on the east side of the plaza features a glass roof constructed over the narrow spaces of two parallel structures. It is divided into four zones in accordance with disaster-prevention and construction programmes. These articulated domains have separate, second facades.

The Research Centre of Advanced Science Technology on the west side of the plaza is organised around courtyards. One block has already been completed. The master plan calls for a second block with the same form. The articulated courtyards of the two blocks will be spatially continuous.

The University of Tokyo is developing a new, separate campus in a suburb, but it is rebuilding the central Komaba II campus at the same time, in the belief that cities and universities need to maintain a close relationship.

South facade: Institute of Industrial Science

Opposite: The path to the east gate from the 'University Plaza'

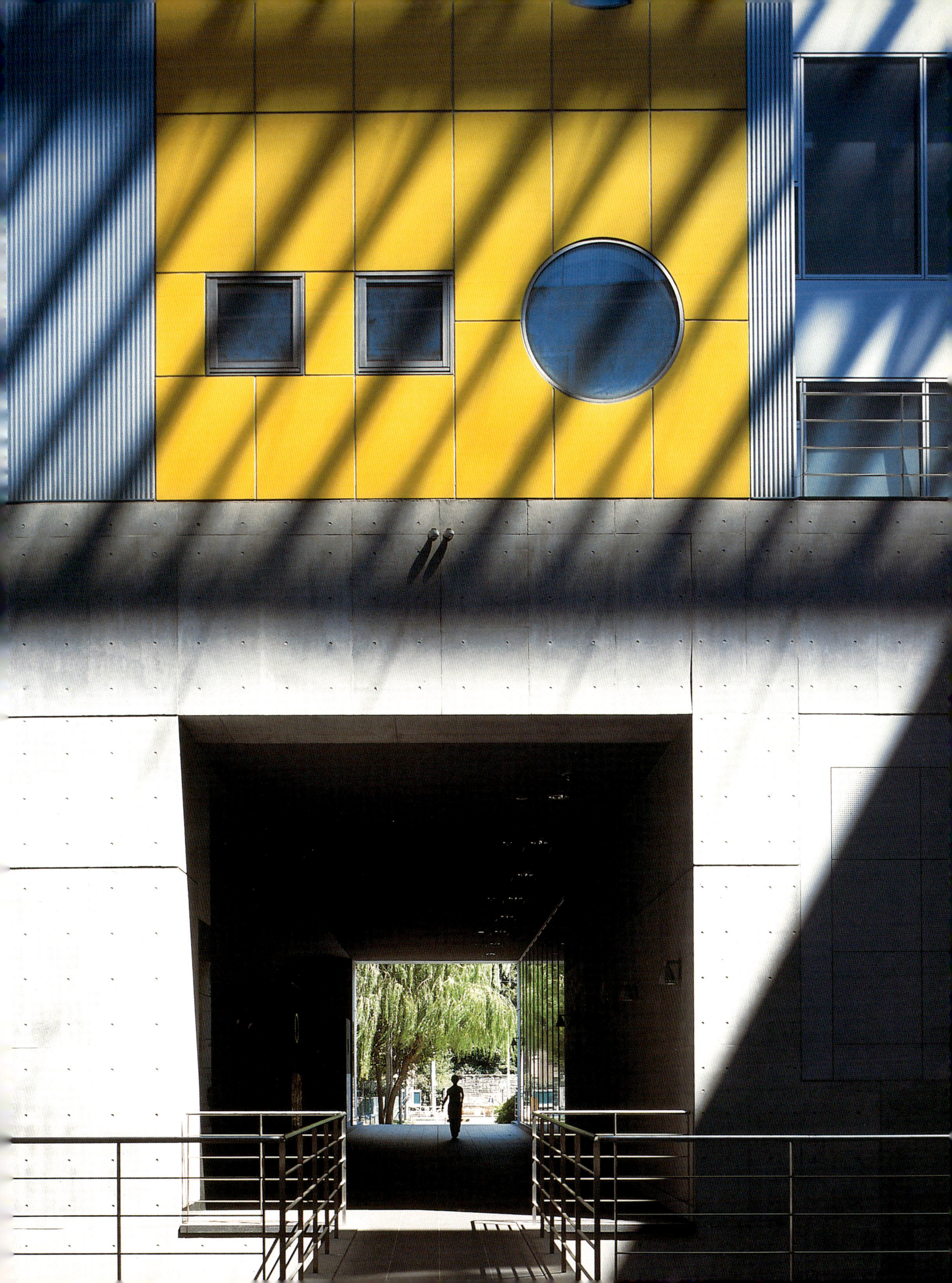

Hiroshi Hara

Above: General view from the west, Institute of Industrial Science

West elevation in the atrium

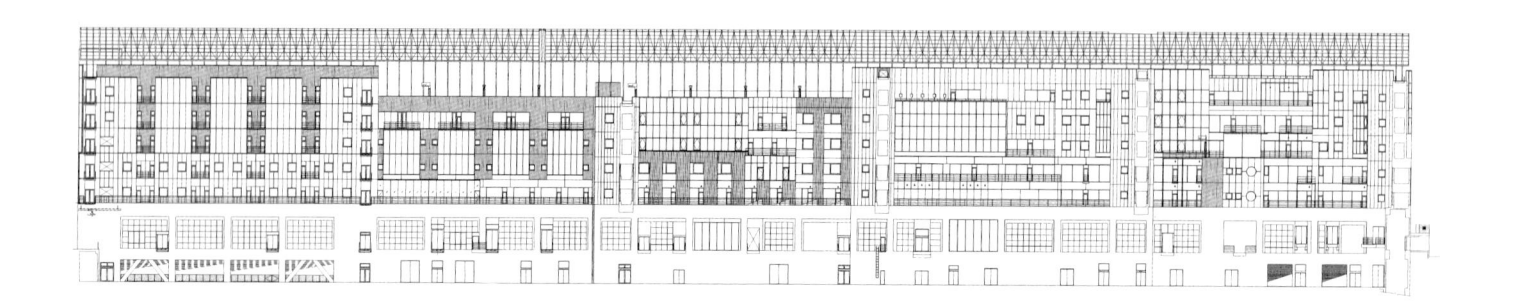

University of Tokyo Komaba II Campus

Overleaf: Upward view of the atrium in the D wing, Institute of Industrial Science

First floor plan

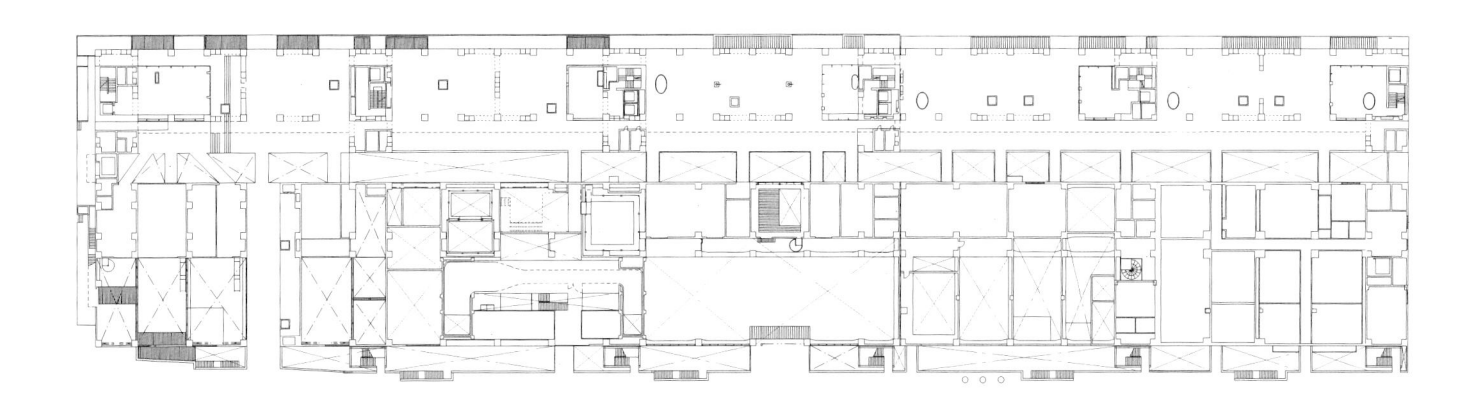

233

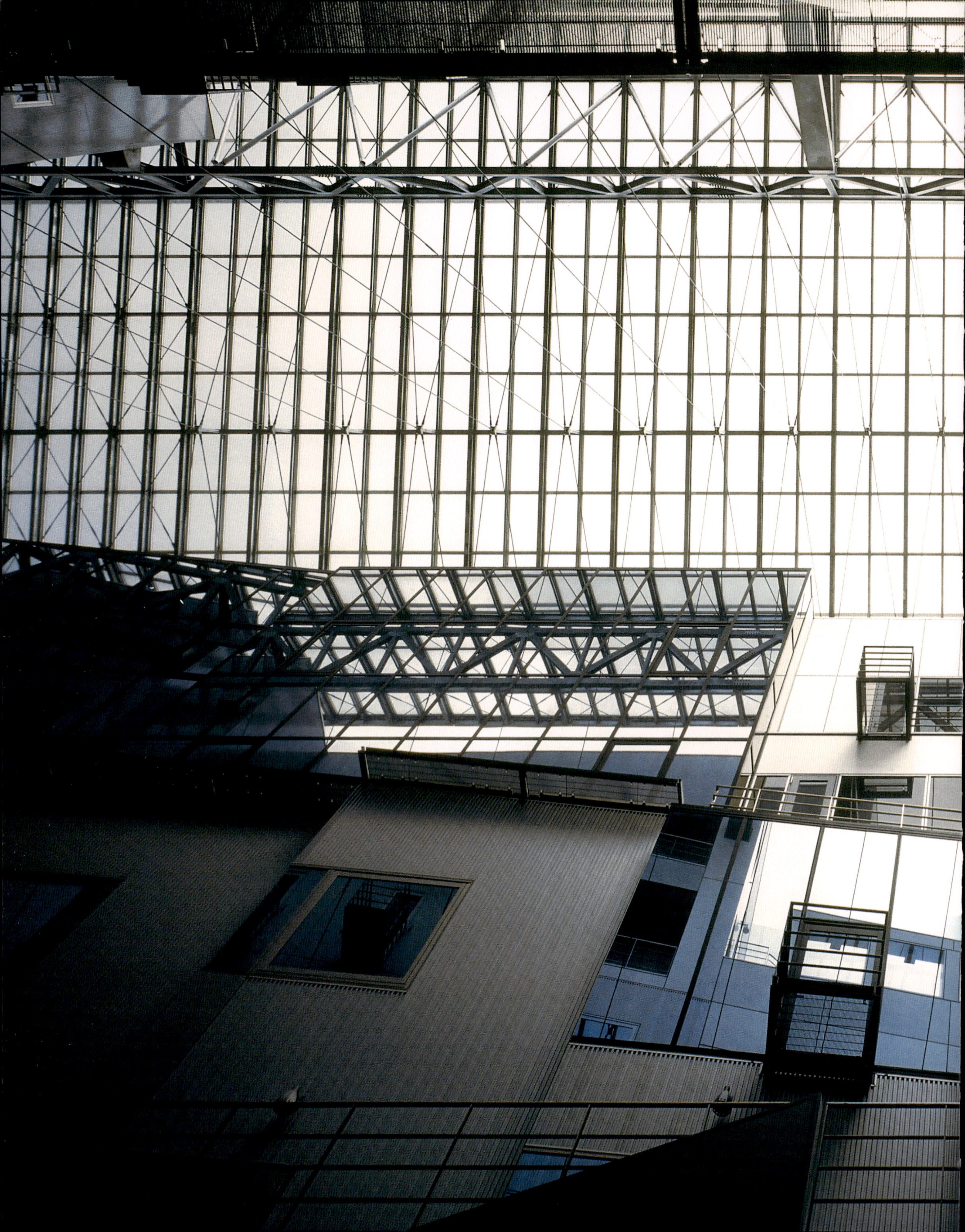

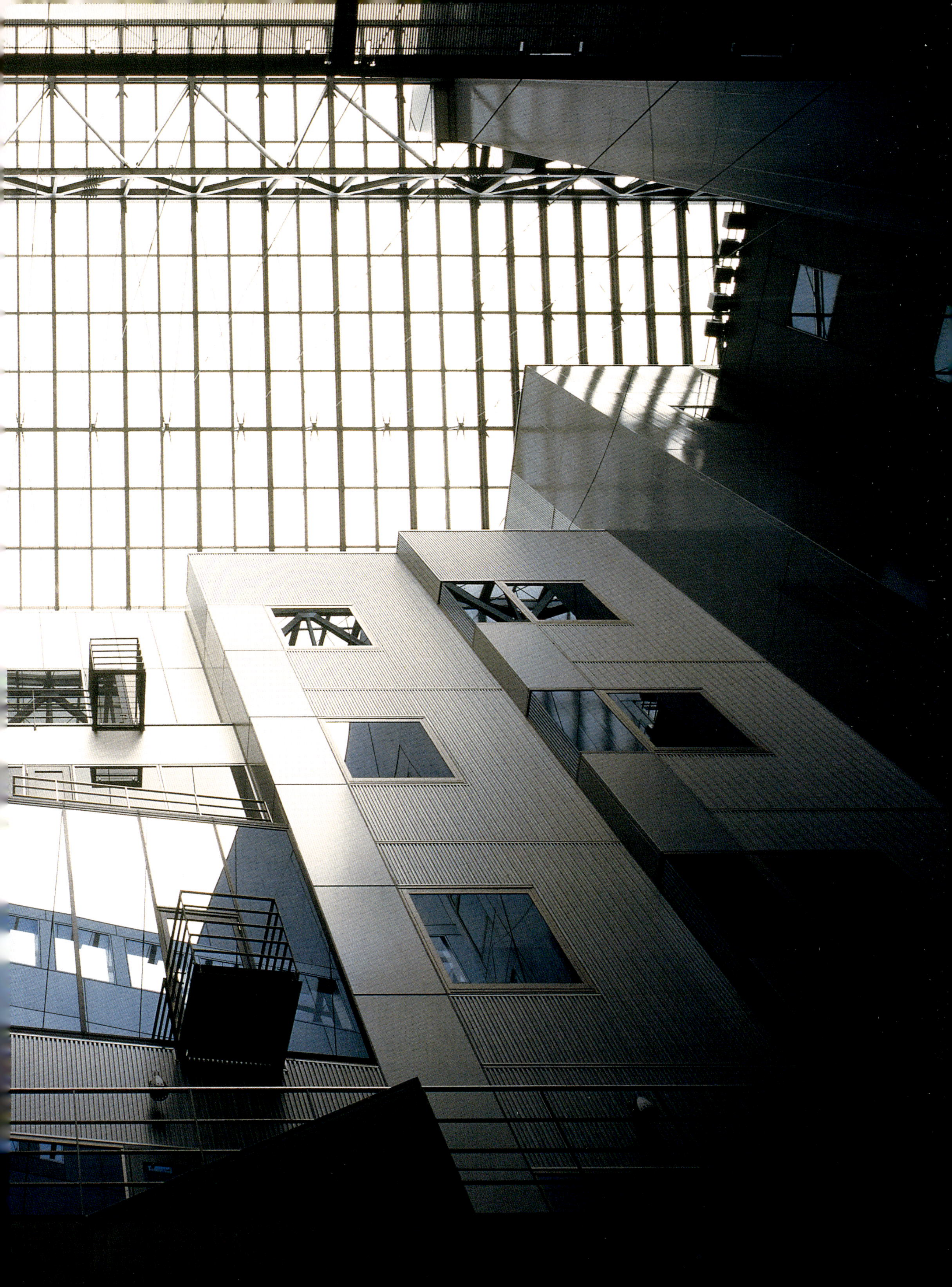

Hiroshi Hara

View towards the high open space, an interior space with a glass roof, seen from the north side of the atrium in the D wing, Institute of Industrial Science

University of Tokyo Komaba II Campus

Upward view of the atrium in the C wing, Institute of Industrial Science

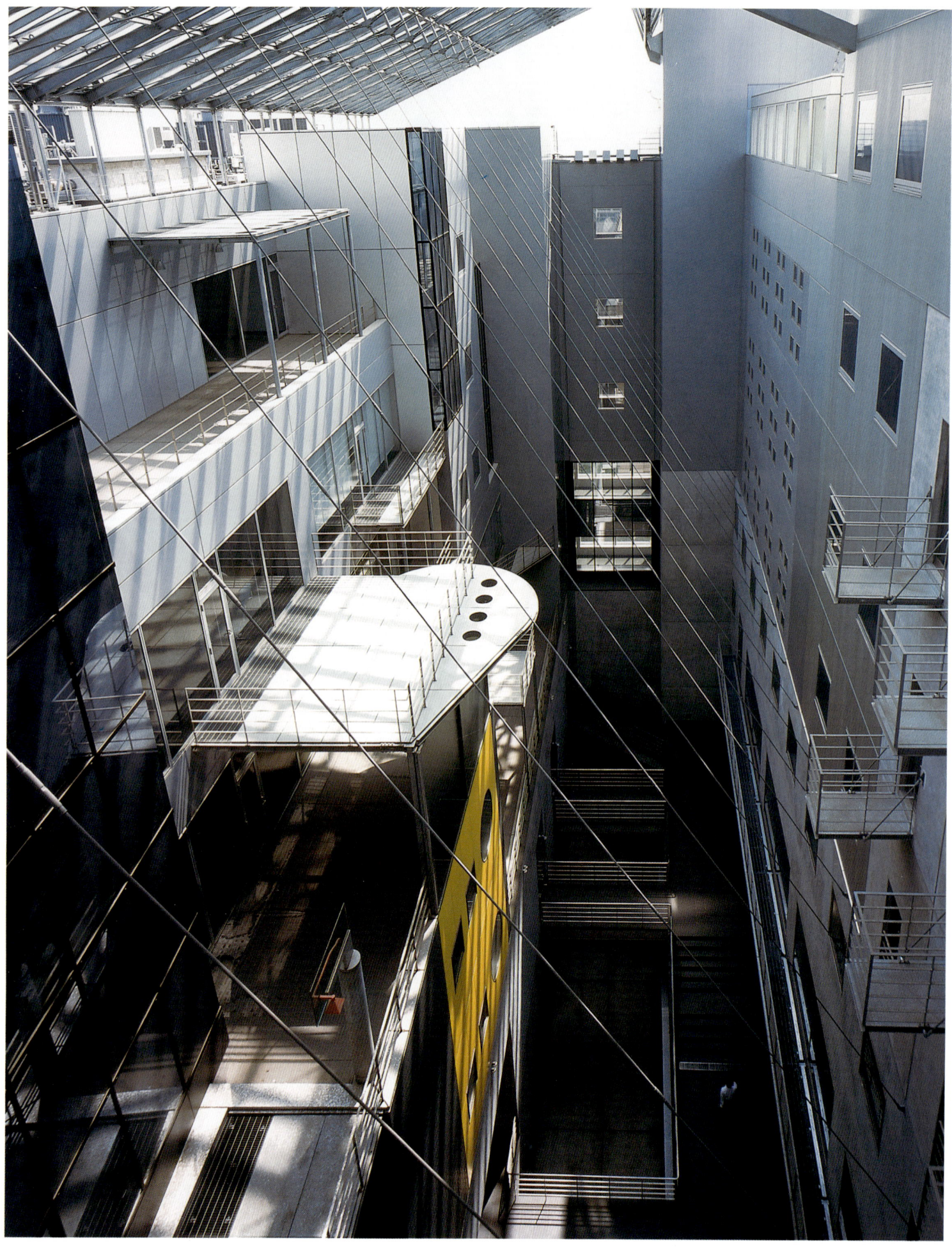

Hiroshi Hara

University of Tokyo Komaba II Campus

Perspectives of the atrium in the B wing, Institute of Industrial Science

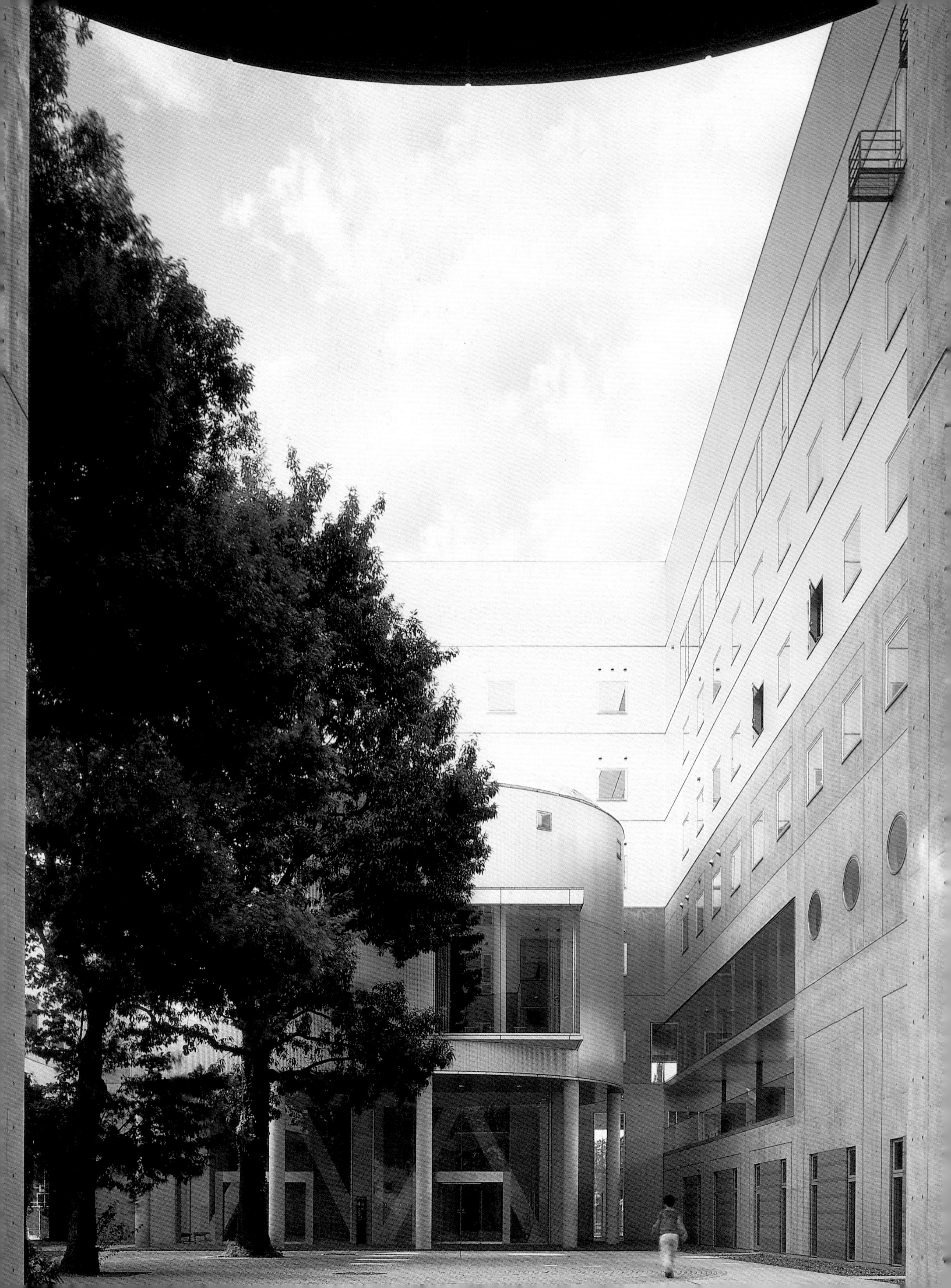

University of Tokyo Komaba II Campus

View from the courtyard from the east, Research Centre of Advanced Science Technology

View from the courtyard from the north, Research Centre of Advanced Science Technology

241

7. Modality / Possibility of Events
by Hiroshi Hara

To me, modality is the most significant concept in architecture. When based on modality, architecture-as-event is realised. Modality is another word for space-time existence: it defines our state of being. Many philosophers and logicians from Aristotle on have considered this matter. I myself come close to the standpoint taken by Spinoza (1632–77). However, while he speculated that there is an invariable substance from which all variant existence comes through modification, I believe that there is only modification concerning nature. In Japanese culture, influenced by Eastern philosophy, modality is called *So*.

In the early days of modernist architecture, the fact that architecture represents space-time existence was consistently emphasised, by Sigfried Gideon, for example. Later, in Mies van der Rohe's universal space, time was substituted by flexibility of space. Later, Team Ten and the Japanese Metabolists interpreted time in their own different ways. But, generally speaking, architectural concepts concerning time have been close to frozen.

I proposed the concept of modality in 1986 upon completion of the Tasaki Museum. I was simultaneously designing Yamato International, and in both buildings I employed a transitional facade and interior space. The term first appeared in the phrase 'from function to modality'. At the time, I believed that functionalism was based on the aesthetics of the mechanism, like two adjacent gears, and that if two objects must be in direct contact, then this is a limitation. In the age of modern electronics, not only adjacent but distant objects too find correlations, which functionality can no longer handle.

Being relatively unacquainted with the English language, I find the pilot's opening communication, 'Can you read me?' quite beautiful. In the daylight, or at sunset, I imagine many people listening to a voice coming from nowhere asking 'Can you read me?' In response, I wanted to make a building that replied 'Yes, I read you very well'. I always have such correspondence with nature in mind; the correspondence between architecture and nature is one aspect of modality.

In contrast to eternal architecture, there is ephemeral architecture. Of course, temporary architecture is ephemeral, but by saying this I refer to buildings that change their state from time to time. They are also based on ephemerality. According to Buddhism, as the Indian philosopher Nagarjuna (c. AD 100–200) said long ago, eternity and the instant are the same. Eternity can be seen in an instant, and this instant will never come again. Following this thought, architecture that never looks the same twice would be the ideal. The appearance of the facades of the Umeda Sky Building and Kyoto Station stem from this concept.

As we have seen, based on my interest in the temporal, I made the installation *Robot Silhouette*, in which light sources such as sweep neon, cold cathode and LED are distributed on twenty acrylic plates. The light sources change every tenth of a second, forming a kind of orchestra that can play specially composed 'tunes' of light. The diagram shows part of the thirteen-minute tune that was played when it was exhibited in the United States. The device generates complicated images through the overlay of lights and illustrations printed on the

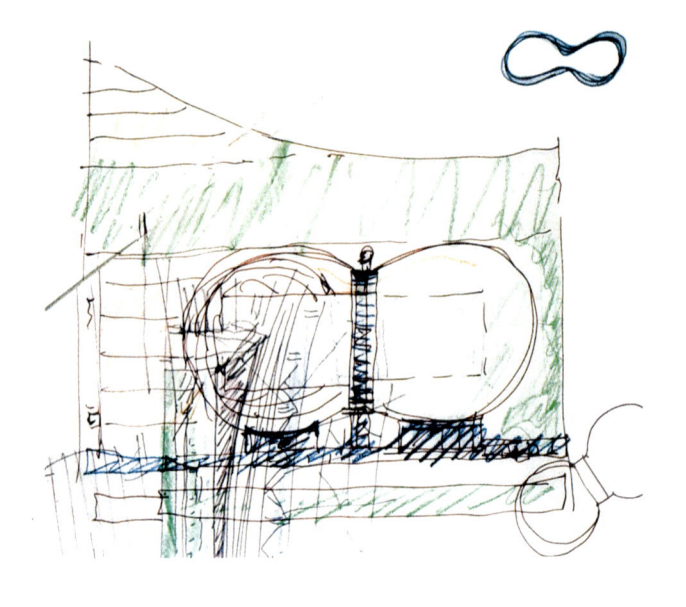

Sketch of Sapporo Dome: plan of dual arena

acrylic plates. The images generated are in this way similar to those in our memory. We can concentrate on a certain part, but the rest is ambiguous. People looking at this orchestra of light are looking into their own consciousness.

A second work, *25 Music Stands*, was made for the Louisiana Museum in Copenhagen. Technically, it is similar to the previous device. But while *Robot Silhouette* generated images as continuum using a multilayered structure, the new device generates individuum images with a tune lasting for about a minute. The diagram depicts the composition and its detail. This is a presentation of a semiotic field, changing from time to time.

Though not as clearly as in these two devices, it is still possible to design changing facades for buildings. However, the concept of modality should directly relate to people's activities. In the Japanese tradition, there is an architectural concept called *shitsurai* (equipment). The Japanese house basically consists of floor, pillars and roof, together with moveable partitions called *fusuma* and *shoji*. Different *shitsurai*s are brought in according to the season and the house is complete. It is possible to consider *fusuma* and *shoji*, or even the floor covering *tatami*, as *shitsurai*s. The winter equipment is represented by *kotatsu*, which is a heater, and the summer by *sudare*, a bamboo-reed blind. The *shitsurai* is therefore a generalised mobile system.

I have always dreamt of architecture that has *shitsurai*s. Architecture with this kind of mobile system is architecture-as-event, and symbolises the concept of modality. I finally realised this at Sapporo Dome (2001), a facility for sports and events.

Sapporo is a city in Hokkaido, the northernmost island of Japan. In order to maintain a turf football pitch, four hours of direct daylight are required, but the weather in this area is cold and snowy. There are two possible mechanical solutions. One is to move the roof and the other is to move the pitch itself outside. The former was almost impossible to implement in Sapporo's climate, so the city chose the latter solution. During the competition, I came up with the idea of a 'dual arena': there would be two adjacent arenas, one covered and the other exposed. The football field travels back and forth between the two.

Sapporo Dome has three formations, or modes:
1. Football Mode
2. Baseball Mode
3. Other Events Mode

These modes are implemented using the following mechanisms:
   a. A hovering field, which floats by compressed air and rotates between the arenas.
   bi. 5,000 moveable seats, that can be folded and stowed away.
   bii. Two wings of moveable stands, each with 1,000 seats, that rotate along the arena walls.
   c. A moving wall that stands at the threshold between the open and closed arena, passed through by the hovering field.
   d. An artificial lawn for baseball.
   e. Curtains for all glass surfaces including the skylight.
   f. Advertisement banners.

Sketch of Sapporo Dome: elevation

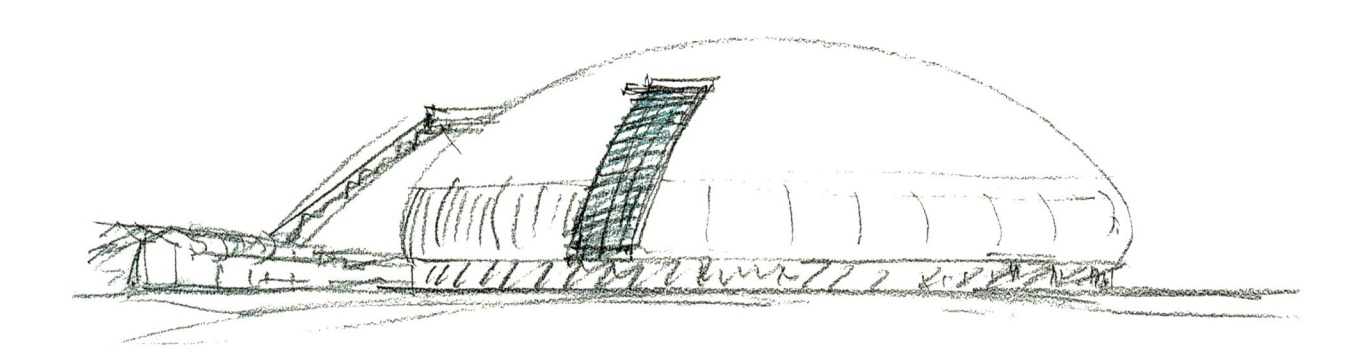

These large-scale *shitsurai*s, a–e, are applied to 1 and 2 to make 3 (usually concerts and exhibitions). There is also an escalator that leads to an observation platform. From here, one can view the arena and the city of Sapporo from above. This would be: '4. Dome Tour Mode'.

Aristotle told us that there are two ways to recognise existence: the real mode (*energeia*) and possible mode (*dynamis*). Built architecture belongs to the real mode, but during the design process to the possible mode. So all architecture should be designed as 'architecture as *dynamis*'. The Sapporo Dome is a unique stadium equipped with moveable devices and at the same time provides an opportunity to reconfirm architecture as *dynamis*.

In the Sapporo Dome, the modes vary from 1 to 4 within the sphere of the possible mode. Together, they compose the modality of the Sapporo Dome.

Architecture should correspond to the transitions of nature. In the case of Sapporo, winter brings a metre of snow and a –10 degree air temperature. We were mostly concerned with the snow on the roof, since its nature is unpredictable when the span is 230 metres; snow always outwits computer simulations. So we had to build and think at the same time. Nonetheless, the stadium lets its modality evolve in accordance with nature's cycle.

The stadium stands on extensive grounds, designed as gardens. The open arena is also a garden, having a football pitch as its main element. Parking areas are planted with trees. Other spaces are covered with grass and lawn. There are twenty small forests characterised by works commissioned from international artists. These attractors are open to the public at all times. The gardens induce unpredictable traversing as discrete semiotic fields.

As long as architecture is event, modality is closely related to the traversing of people. In the Sapporo Dome, when there is an event, there are four involved parties: the organisers, players, spectators and TV audience. In mode 4, a party will visit both the garden and the building. The organiser (actually the people of Sapporo, for it is a public stadium) uses devices a–e and produces urban space. Indeed, the people are the architects who conceptualise and realise modality. This sense of the user being the architect is true for all buildings, but is more evident in Sapporo. The traversing of the organiser is thus the designer of modality.

In certain cases, the stadium turns into a theatre. The audience traverses in a predictable way since it knows where to go. However, if the event is well organised, people can visit the observation platform, sports gallery and restaurants before and after, or perhaps during, the event. So unpredictable traversing is also possible, including within the gardens. There is no special opportunity for the players or the distant TV watchers. To the players, only the spectators exist and the building should be out of mind. As far as the distant audience is concerned, I hope it will be attracted to visit the site in the future.

Sketch of Sapporo Dome: section of closed arena

Modality / Possibility of Events

Sketch of Sapporo Dome: interior of closed arena

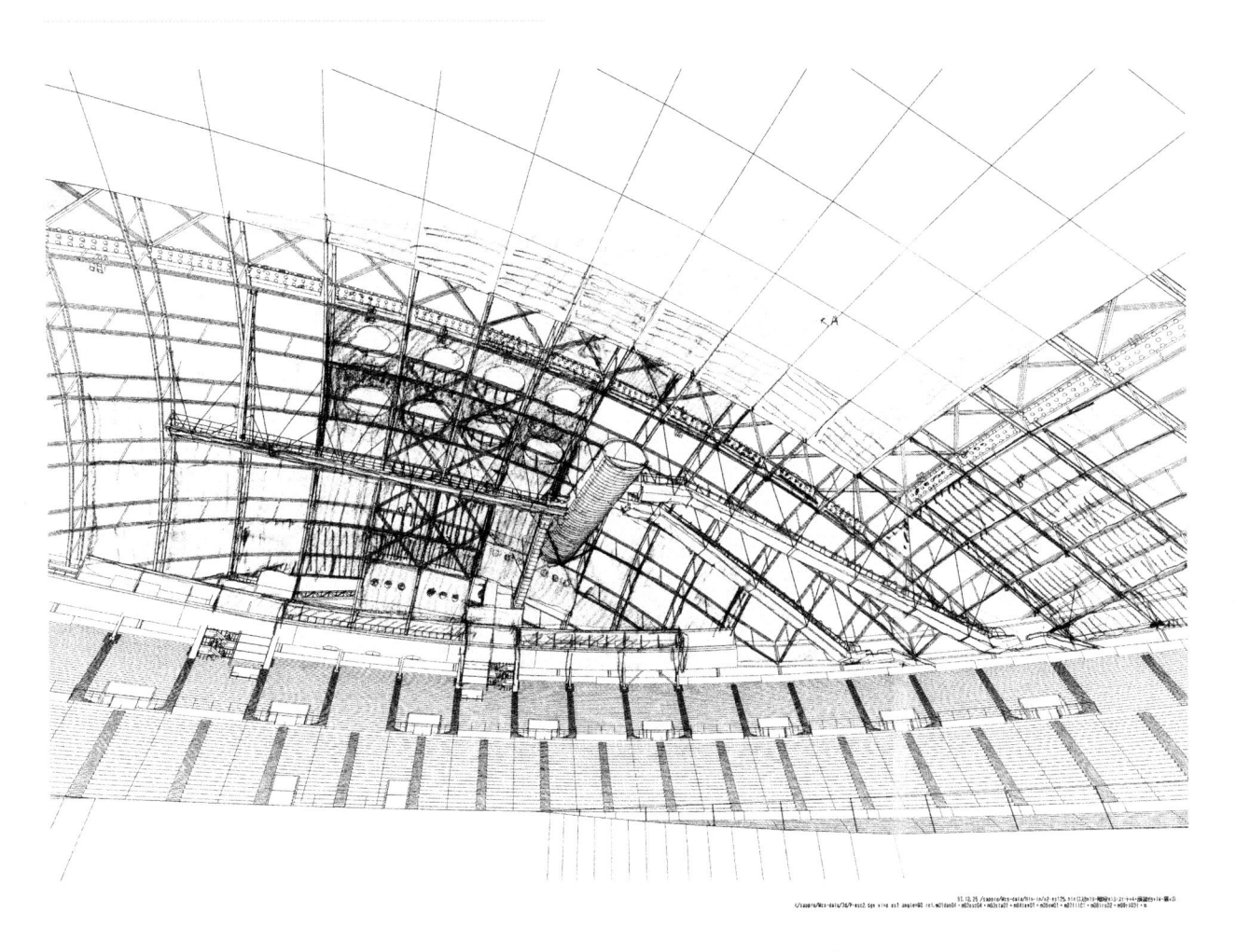

Hiroshi Hara

Modal Space of Consciousness / Robot Silhouette 1984–6

This work is a presentation device with a multilayered structure. Based on an installation exhibited in Graz in 1984, it was created for an exhibition planned by the Walker Art Center in Minneapolis in 1986. The basic unit is a device made of two layered acrylic panels, each measuring 180 x 80 cm. In all, twenty units are used in the optical presentation: four fan-shaped acrylic-panel devices placed on top of cylinders and sixteen acrylic-panel devices placed directly on the floor.

Fragments of the silhouettes of a man and a woman and the design drawing for an automated apparatus used to manufacture watches were inscribed on the acrylic panels of each unit. A fluorescent lamp with a dimmer was concealed under the acrylic panels and lit up the inscriptions on, and the edges of, the panels. In addition, cold-cathode, LEO and sweep-neon lamps above the acrylic panels were turned on and off at one-tenth-of-a-second intervals by computer. There were eight types of light sources in all. The twenty units could be arranged in space in different ways depending on the venue.

An infinite number of 'compositions' can be created through control of the flashing of lights, the movements of the sweep-neon lamps and the dimming of the fluorescent lamps. The 'score' shown here is part of a composition whose performance lasts slightly less than fourteen minutes and involves 8,192 steps.

At each moment the viewer sees a complex image of lights. It is difficult to understand the image in its entirety. Like a picture in our memory, each is a 'fragmentary image'. In this sense, the work is intended to provide a glimpse into our own consciousness.

From the standpoint of multilayered structures, this is an experimental device to see how the whole is formed through the overlapping of the parts and to discover to what extent effects can be deliberately planned. From the standpoint of the concept of modality, it is a model for the simultaneous planning of time and space. In any building, the appearance of the inside and the outside changes as time passes. This can be said to be a device that improves our understanding of the meaning of planning in relation to changes in the appearance of a building or a city through time. *HH*

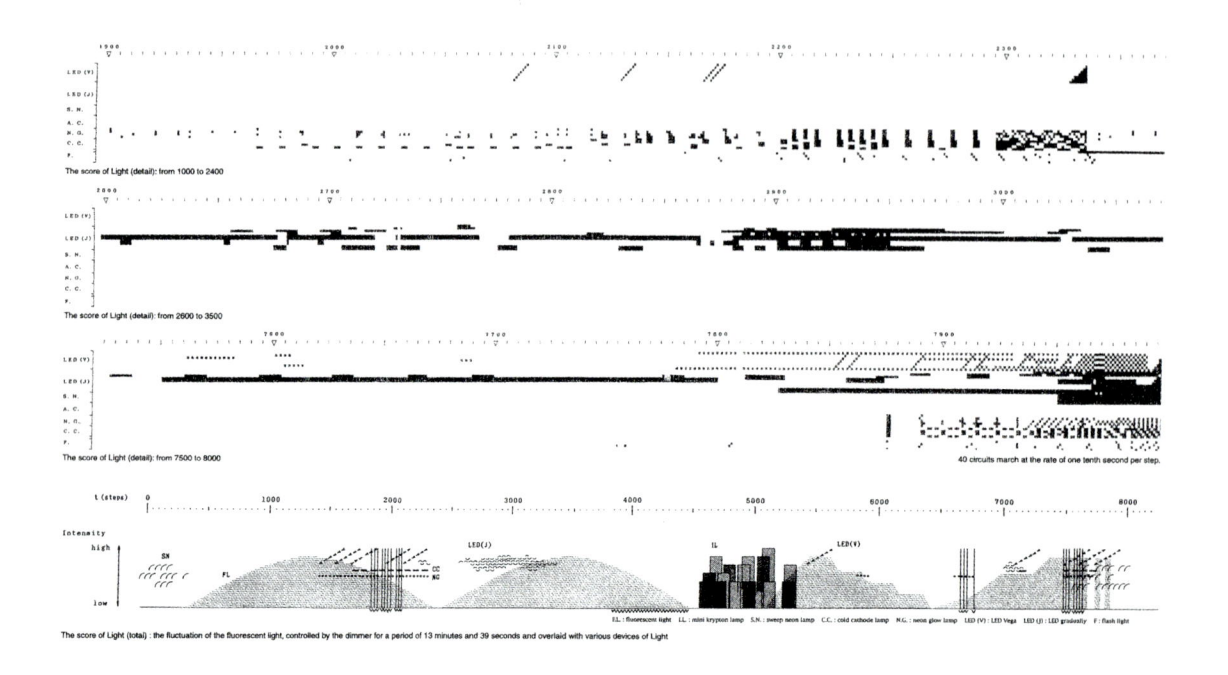

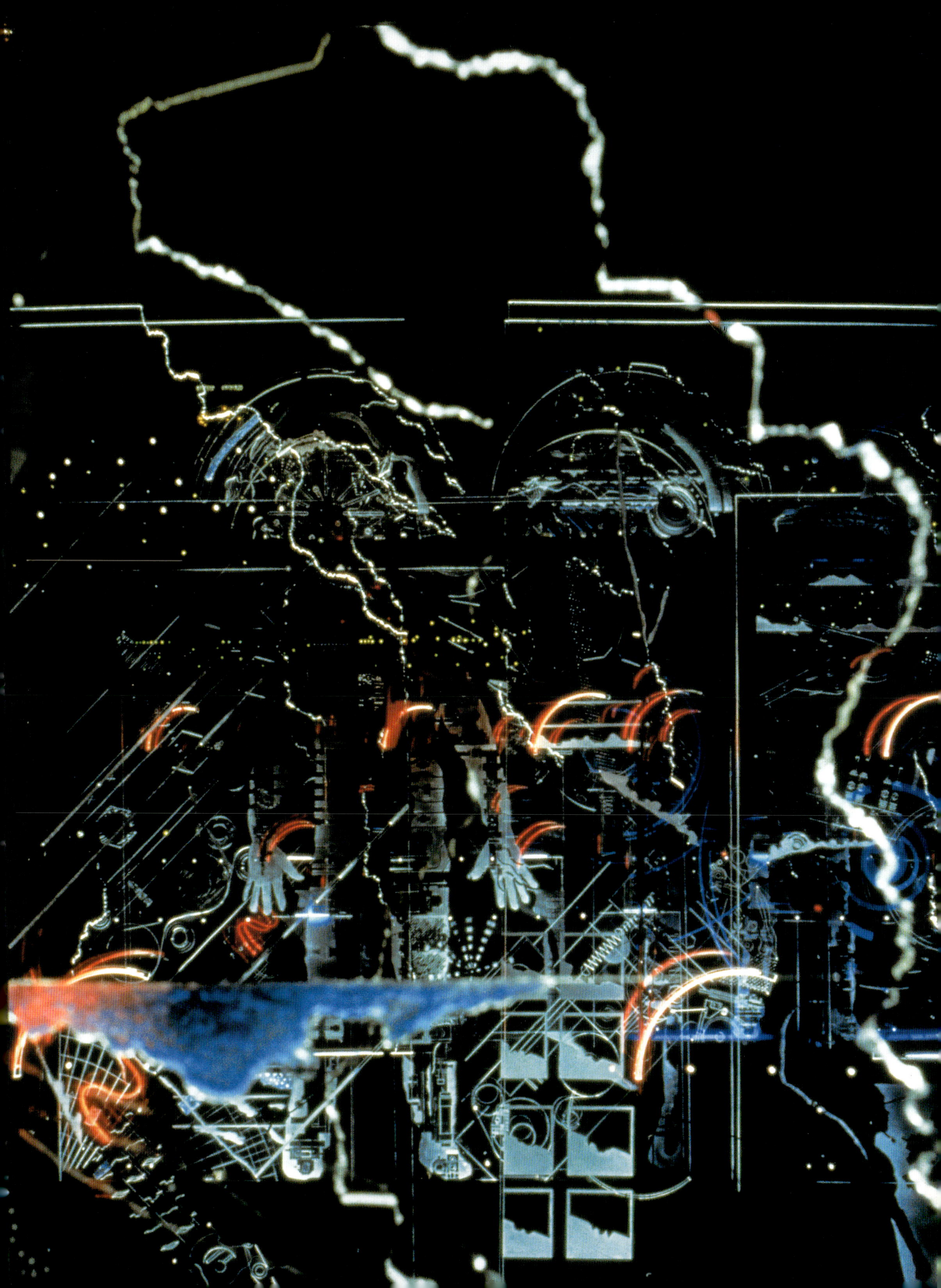

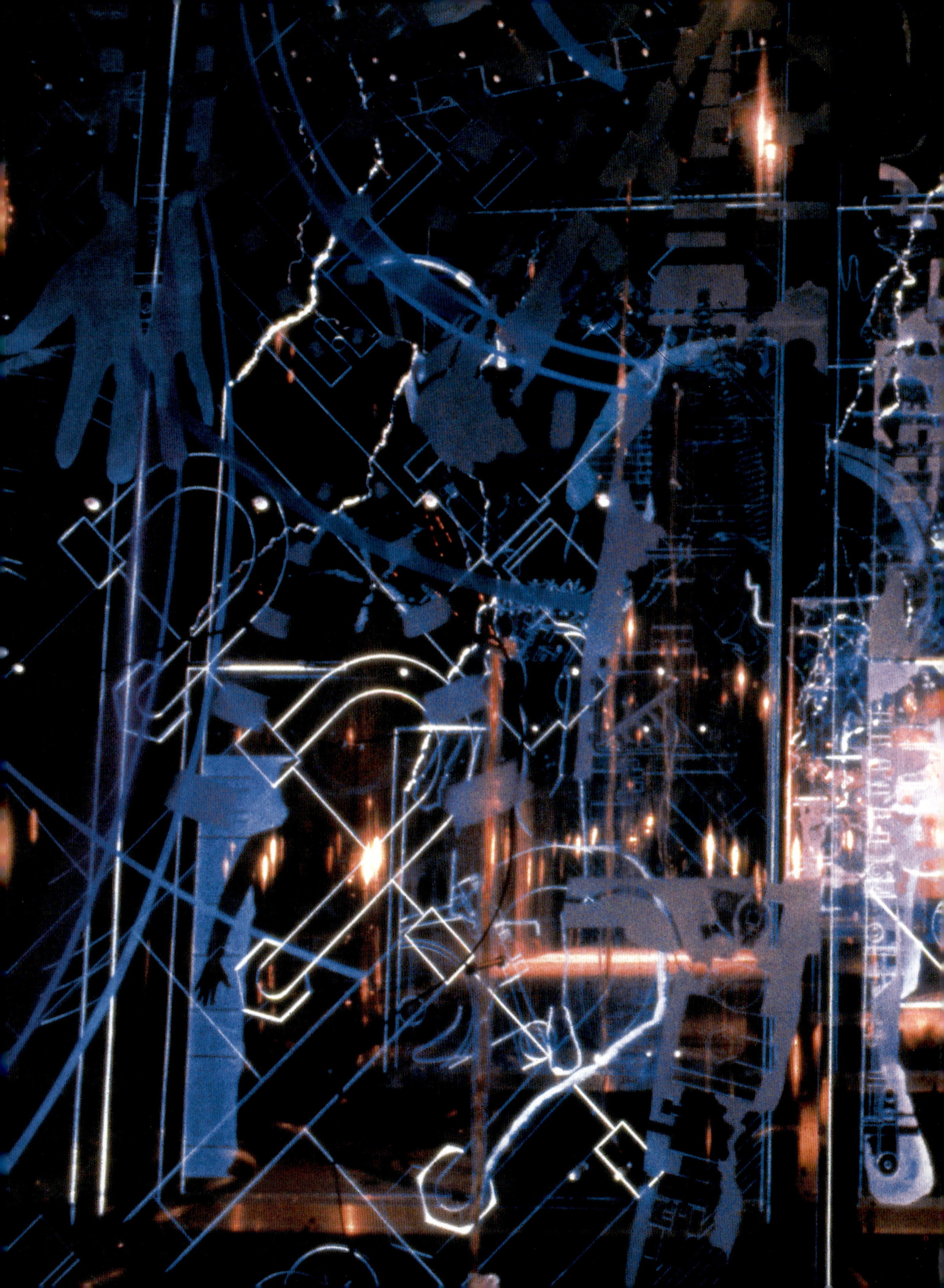

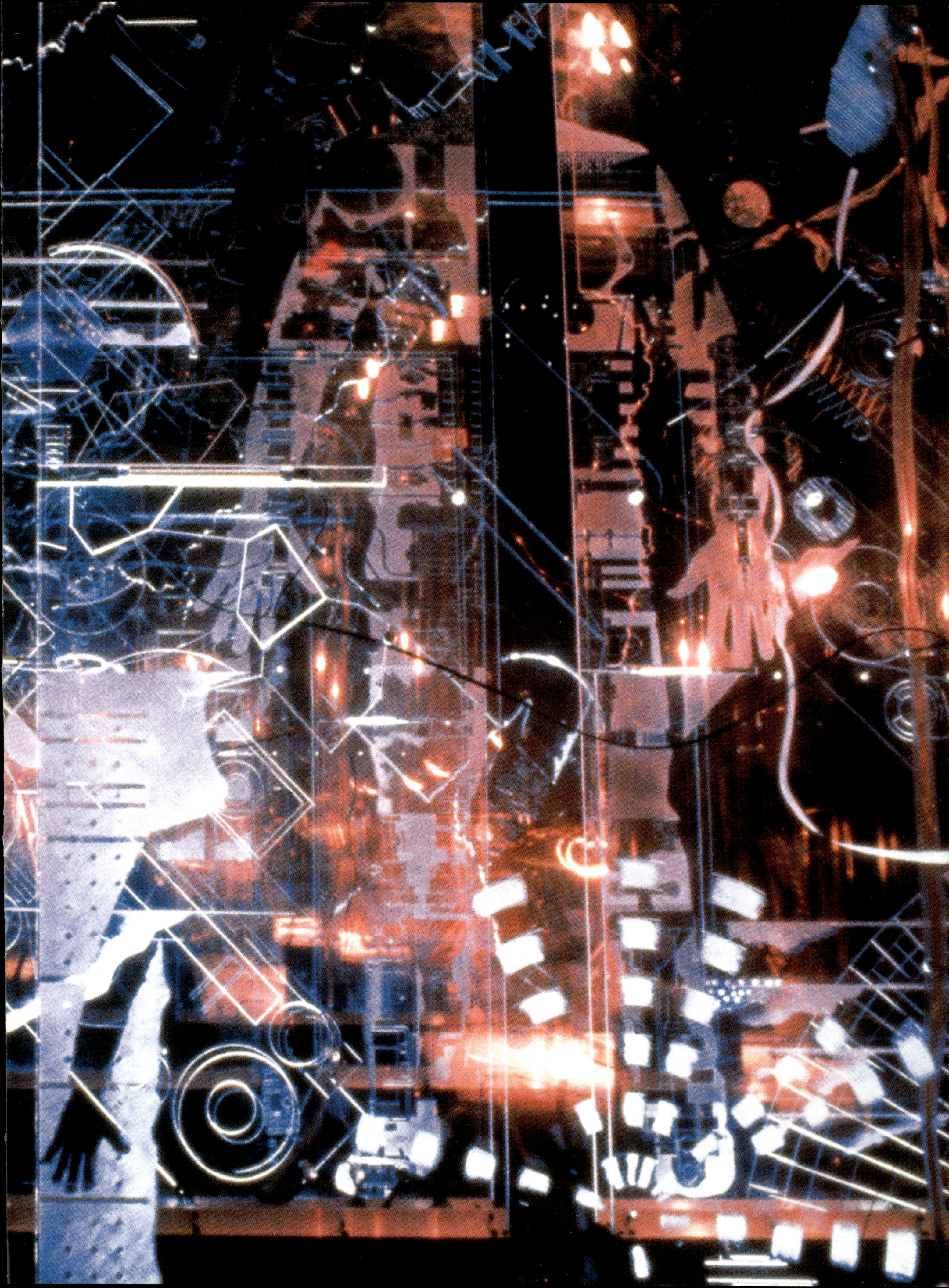

Hiroshi Hara

Twenty-Five Music Stands, 1995

This work, which is based on a similar system to *Robot Silhouette/Modal Space of Consciousness*, was commissioned by the Louisiana Museum in Copenhagen.

While *Robot Silhouette* produces continuous images, this device is intended to produce discrete images. The difference between the two works might be likened to the difference between an agglomerated settlement and a dispersed settlement. The presentation device is completely different from that of *Robot Silhouette*. It consists of various light sources and reflective panels mounted on twenty-five music stands. It is easier to understand than *Robot Silhouette* because each part is illuminated separately, but the overall image at any instant is still difficult to read.

The device is a model for the presentation of a 'discrete semiotic field'. This is composed of a collection of simple elements consisting of points or lines of light and a collection of elements of light with distinctive forms or colours that serve as attractors. If a 'composition of light' is created with the collection of attractors in mind, it becomes relatively simple to plan time and space.

The 'full score' shown here is that of a 'composition' performed in the exhibition that was held in the Louisiana Museum and other venues in Europe. The performance time was eleven minutes and twenty-five seconds.

The device is currently exhibited on the fortieth floor of the Umeda Sky Building. However, the composition it performs there is a short, easily understood one.

A performance, that is, a presentation of lights, can be considered a kind of traversing. The modes of traversing can be divided according to the way in which lights are grouped together. The main classifications are as follows:

ø: a condition in which no light image exists
X1: an image made up of a single light source
X2: a complex image made up of several light sources
X : an image produced by all the light sources

Such an interpretation is possible because this is a 'discrete sign field'. The composition, that is, traversing, can be assembled by appropriately combining {ø, X1, X2, X}.

Temporal and spatial changes in themselves are in accord with the concept of modality. The changes in modality that can take place in a tenth-of-a-second interval, from my experience in composition, are as follows:

0 – duration
1 – gradual appearance of element (dawn modality)
2 – gradual disappearance of element (twilight modality)
3 – sudden, unforeseeable change.

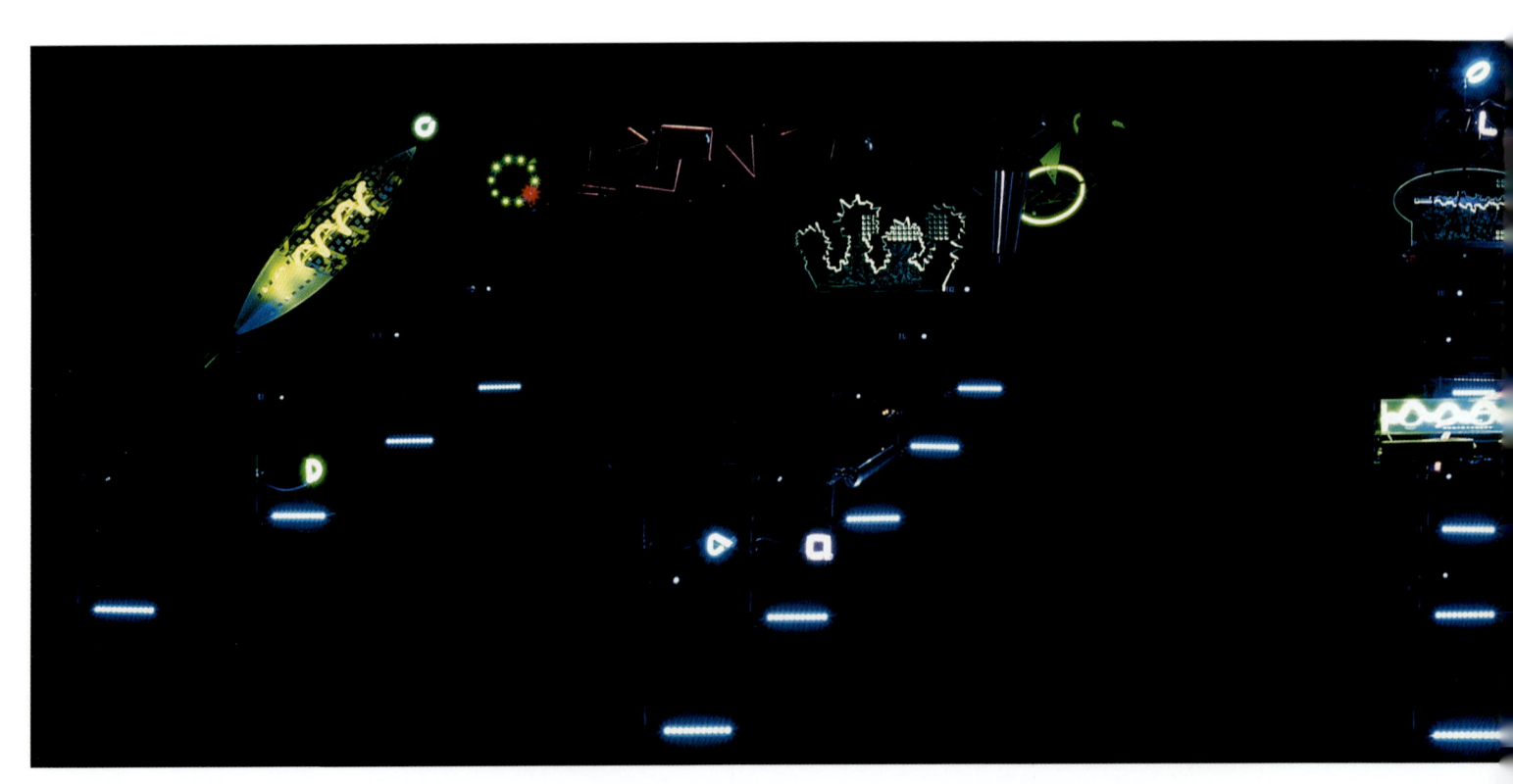

Twenty-five Music Stands

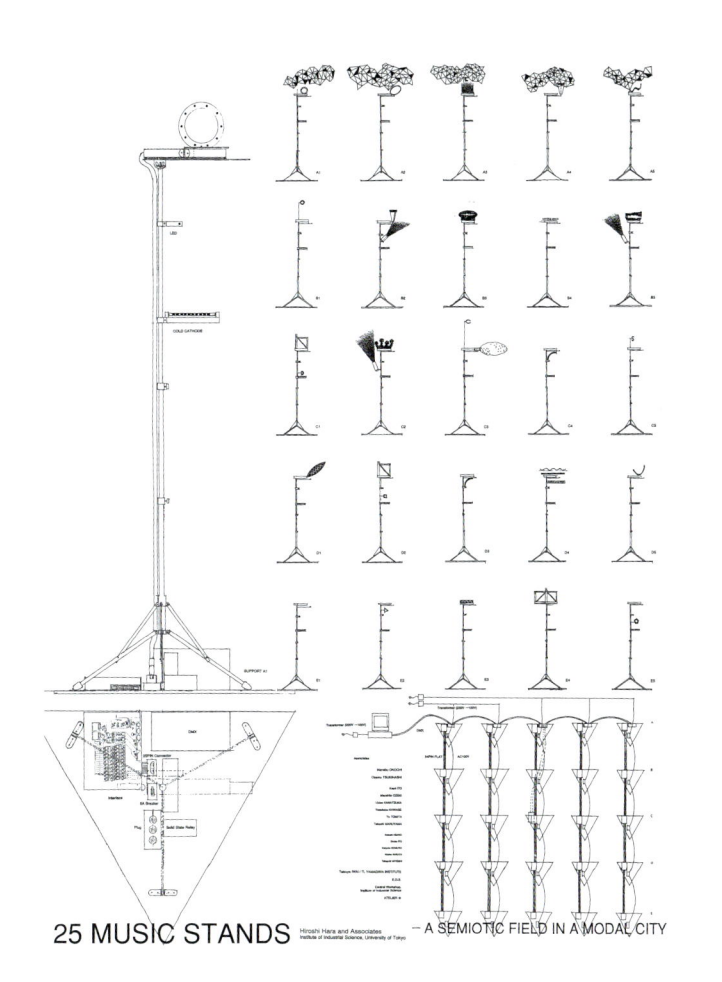

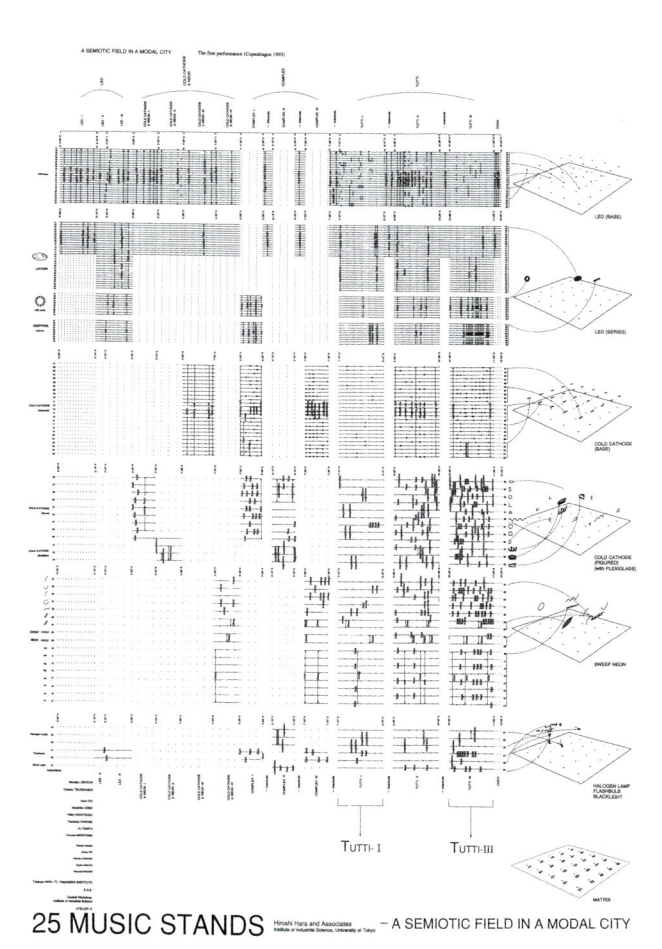

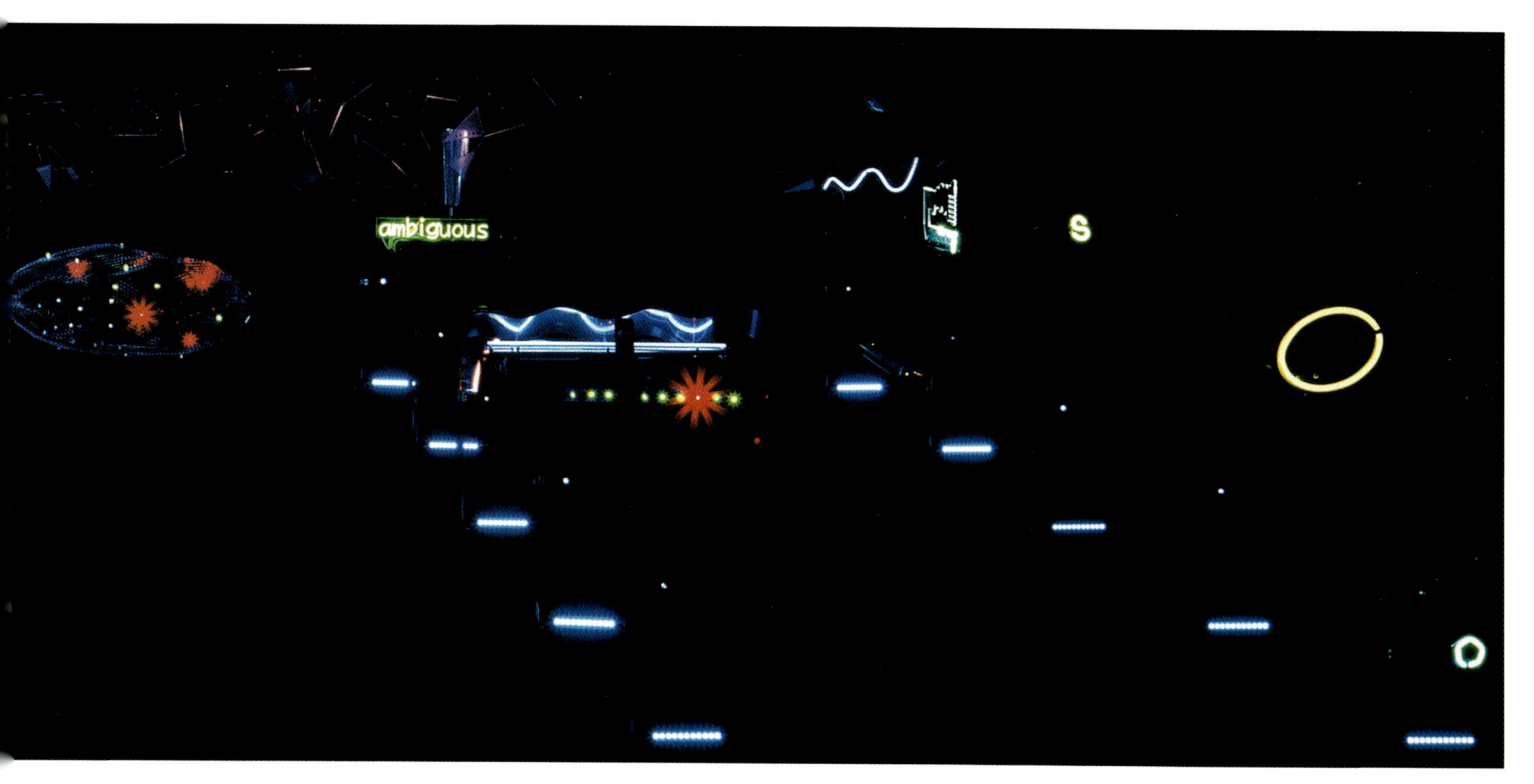

25 MUSIC STANDS — A SEMIOTIC FIELD IN A MODAL CITY

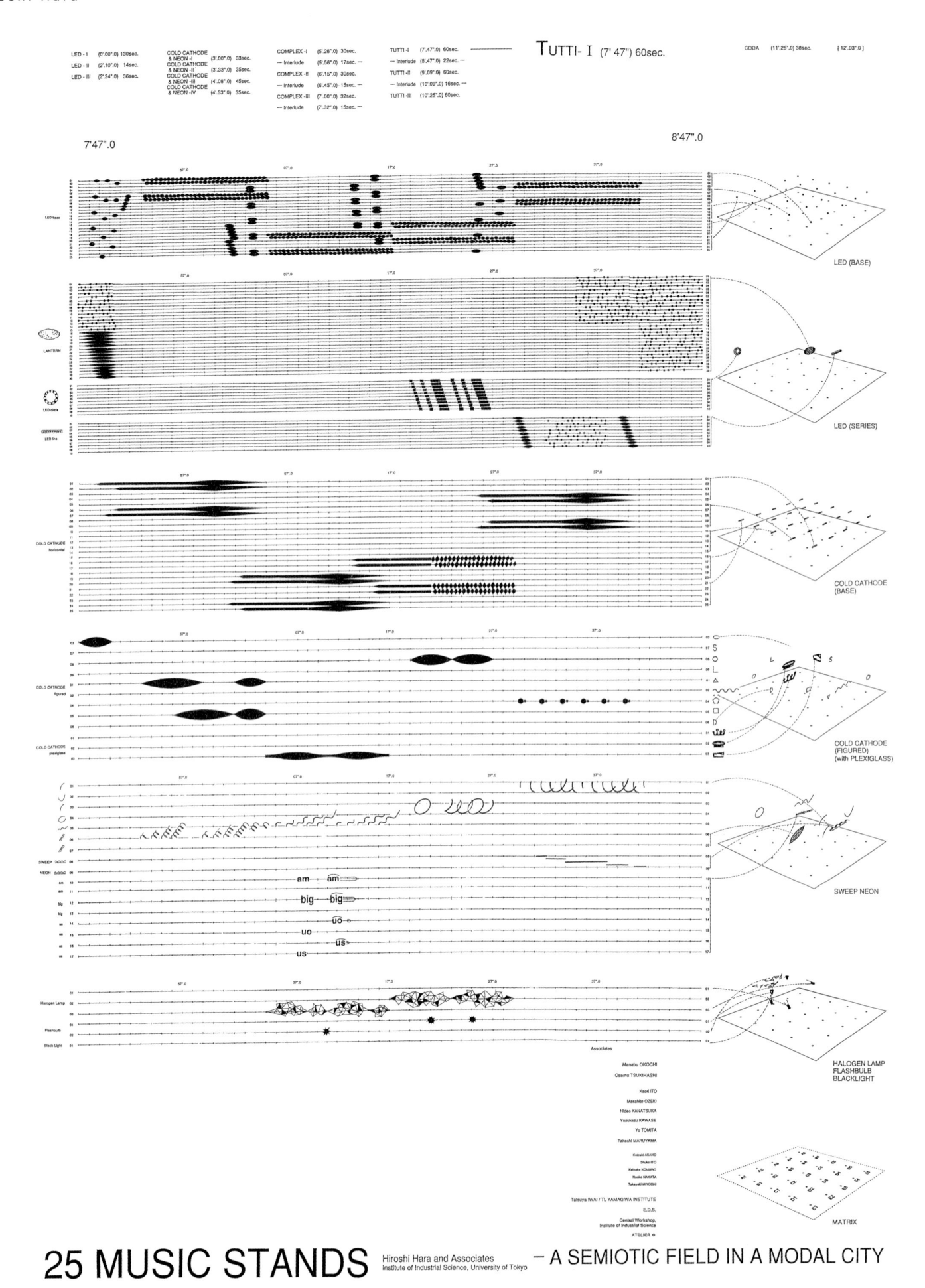

Hiroshi Hara

25 MUSIC STANDS — A SEMIOTIC FIELD IN A MODAL CITY

Hiroshi Hara and Associates
Institute of Industrial Science, University of Tokyo

25 MUSIC STANDS — A SEMIOTIC FIELD IN A MODAL CITY

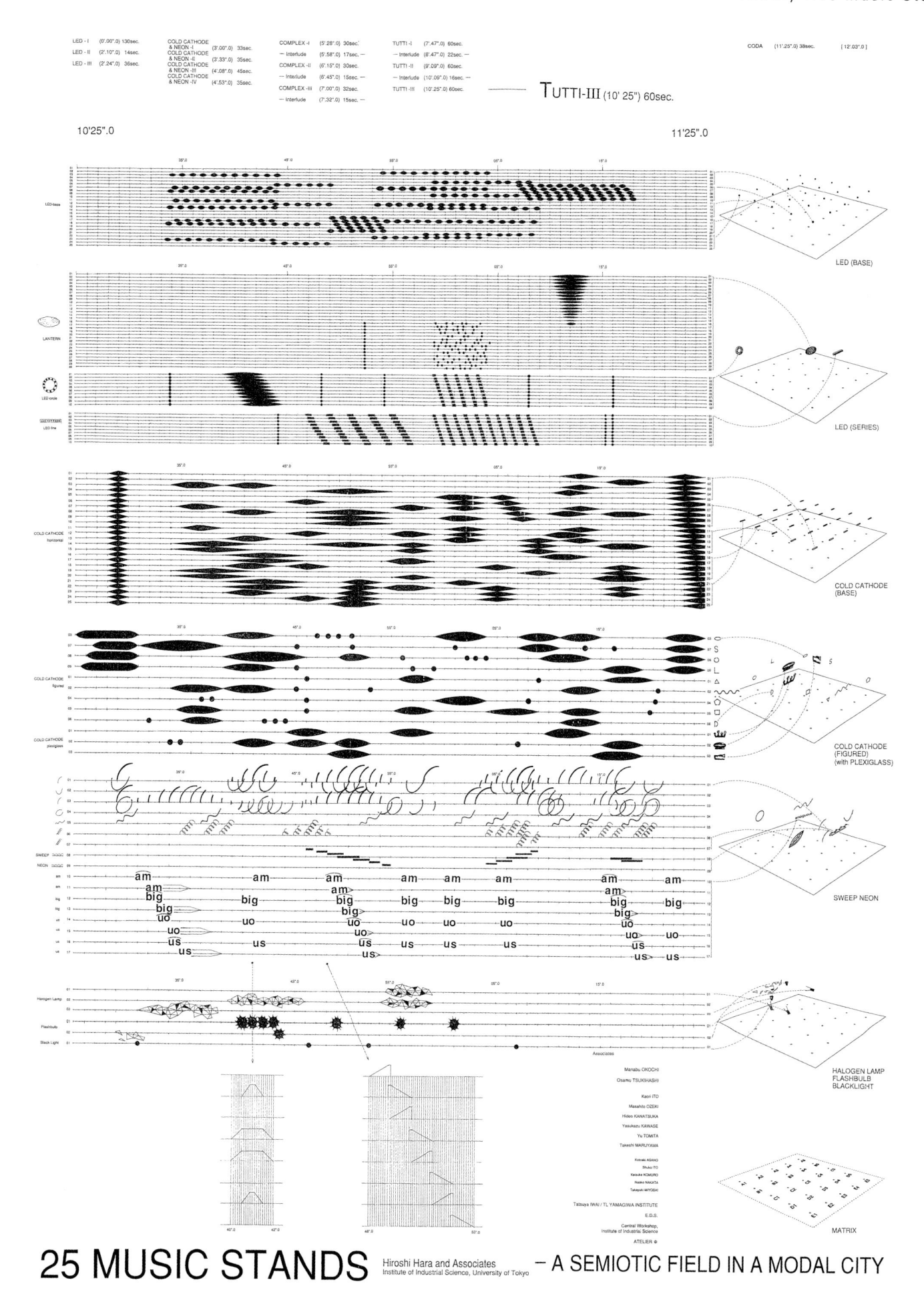

Hiroshi Hara and Associates
Institute of Industrial Science, University of Tokyo

Twenty-five Music Stands

Hiroshi Hara

Sapporo Dome, Sapporo, Hokkaido, 2001

Sapporo is Japan's northernmost major city, with a population of 1.8 million and a latitude of 43 degrees North. Sapporo hosted the 1972 winter Olympics, and has been selected as one of the ten cities to host the Football World Cup in 2002. At the time of selection, the City of Sapporo had been preparing to construct an all-weather dome capable of handling World Cup football, with a convenient turf-changing system. An international competition was held in 1996 for proposals on both the design and the technology, and the Hiroshi Hara Group's plan was selected as the best among worldwide contenders. The silvery egg-shaped dome will appear on Sapporo's scenic Hitsujigaoka in 2001.

The Sapporo Dome has two major characteristics. The first stems from the need to have a natural grass field within a dome for the World Cup. To satisfy this technical need, a 'dual arena' shape was designed, with open and closed arenas facing each other. The football field is laid on what is also called a 'hovering stage', which can be moved between the two by being lightened with a cushion of air and transported on wheels. This mechanism allows for an all-weather football field with a pitch that is always in good condition. In addition, the two arenas can be transformed to adapt to the most appropriate mode for a given event, such as baseball games, concerts and so forth within five hours.

The dome's other characteristic is the gradual transformation of present farmland into a 'sports garden' through a landscaping policy called 'gardening'. The site will be divided into five botanical strips, parallel with the boulevard. Each strip will have its own unique nature. Further, a 'compound' of trees surrounds the entire site, with walkways provided through the greenery. These will be ever-changing and ever-growing components of the sports garden, and will provide a variety of scenery around the dome. This nature-friendly method also allows for a major dome to be built with minor environmental effects on the present residential areas and the Hitsujigaoka district.

Even after this innovative system presents an exciting stage for the World Cup in 2002, the Sapporo Dome with its dual arenas and hovering field will continue to serve as a catalyst for advanced sports activities as well as for various types of events on the public level. The gardening carried out here will provide a new experience for the people of the north, through the overwhelming scenery surrounding the dome and the magnitude of the garden itself. For these visitors, an amusement complex with escalators, an observatory and a small shop will be available. *HH*

General view from the west

Opposite: 'Hovering stage' moving from open arena to closed arena
Overleaf: View of 'Clopen' shell and town seen from the northeast

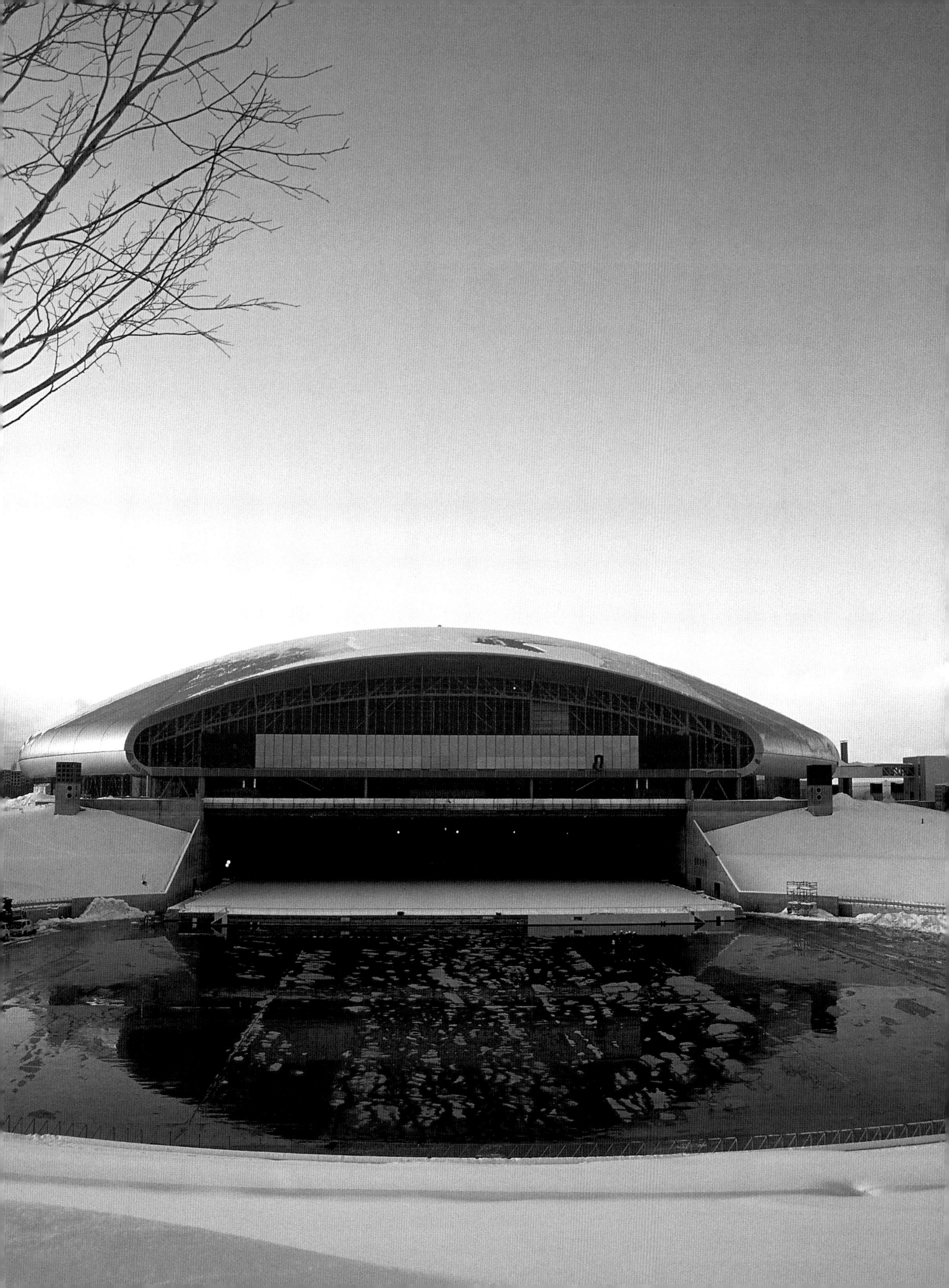

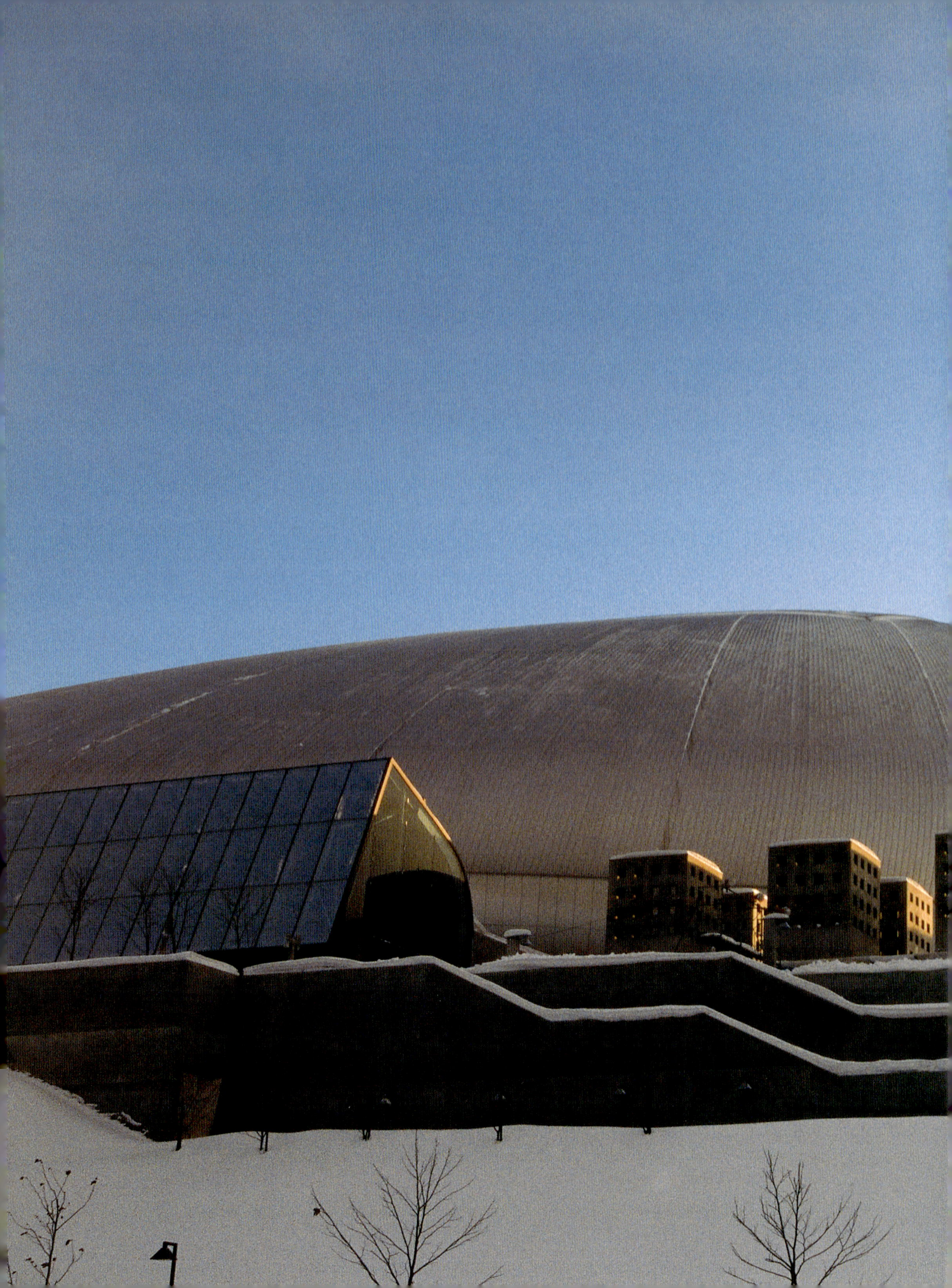

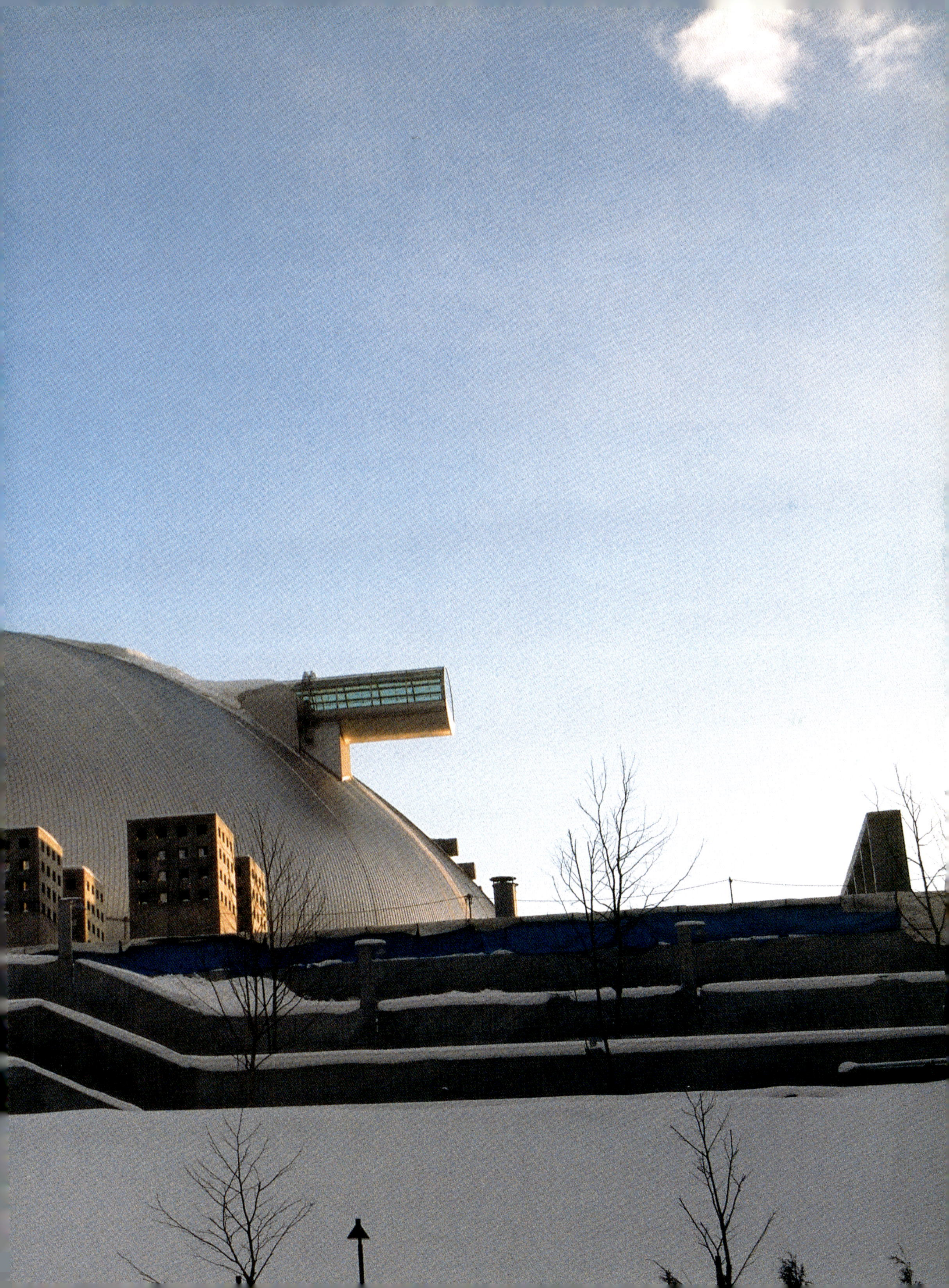

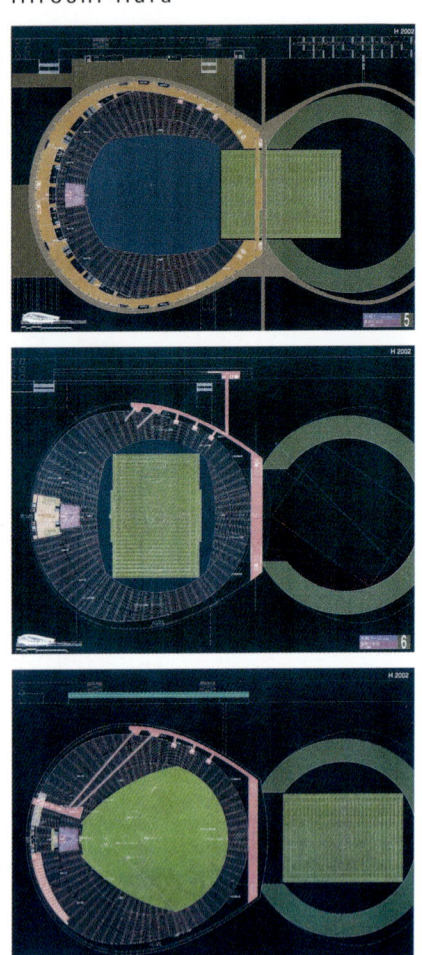

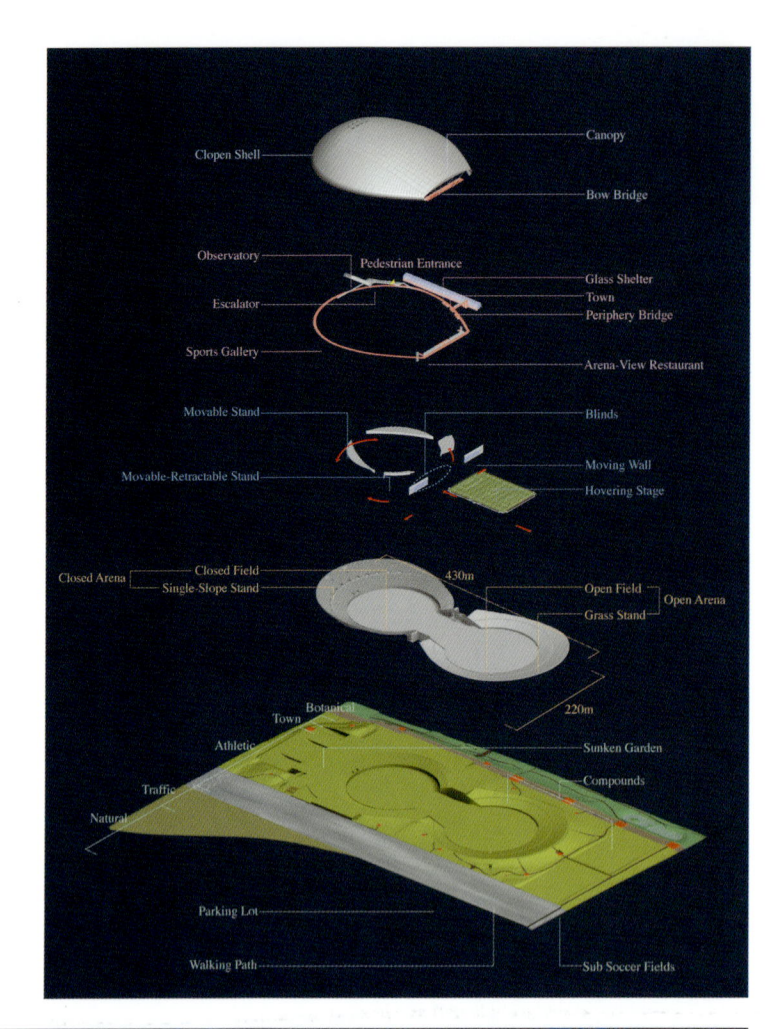

Hiroshi Hara

Sapporo Dome

259

PH Studio

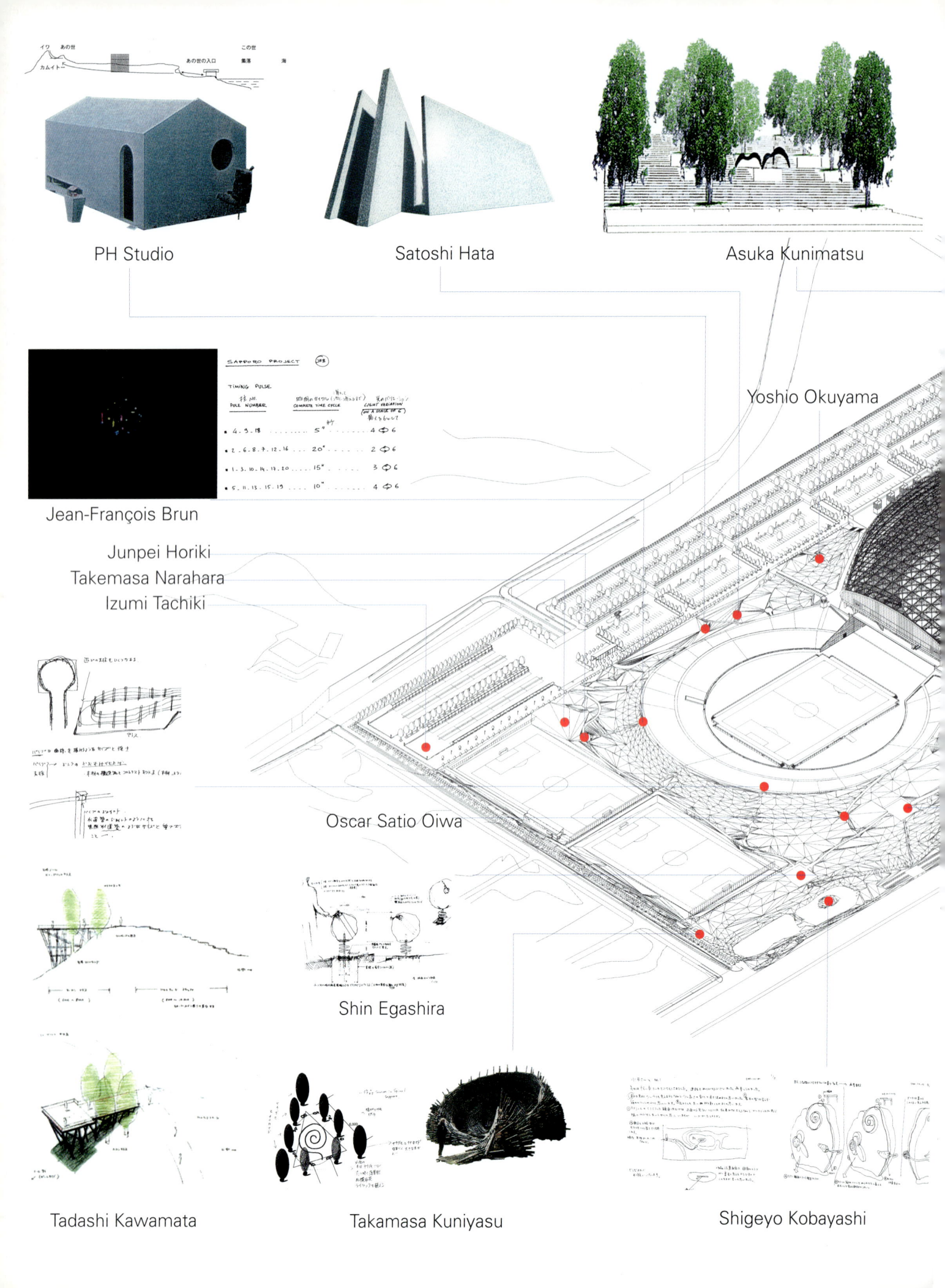

Satoshi Hata

Asuka Kunimatsu

Jean-François Brun

Yoshio Okuyama

Junpei Horiki
Takemasa Narahara
Izumi Tachiki

Oscar Satio Oiwa

Shin Egashira

Tadashi Kawamata

Takamasa Kuniyasu

Shigeyo Kobayashi

Kan Yasuda

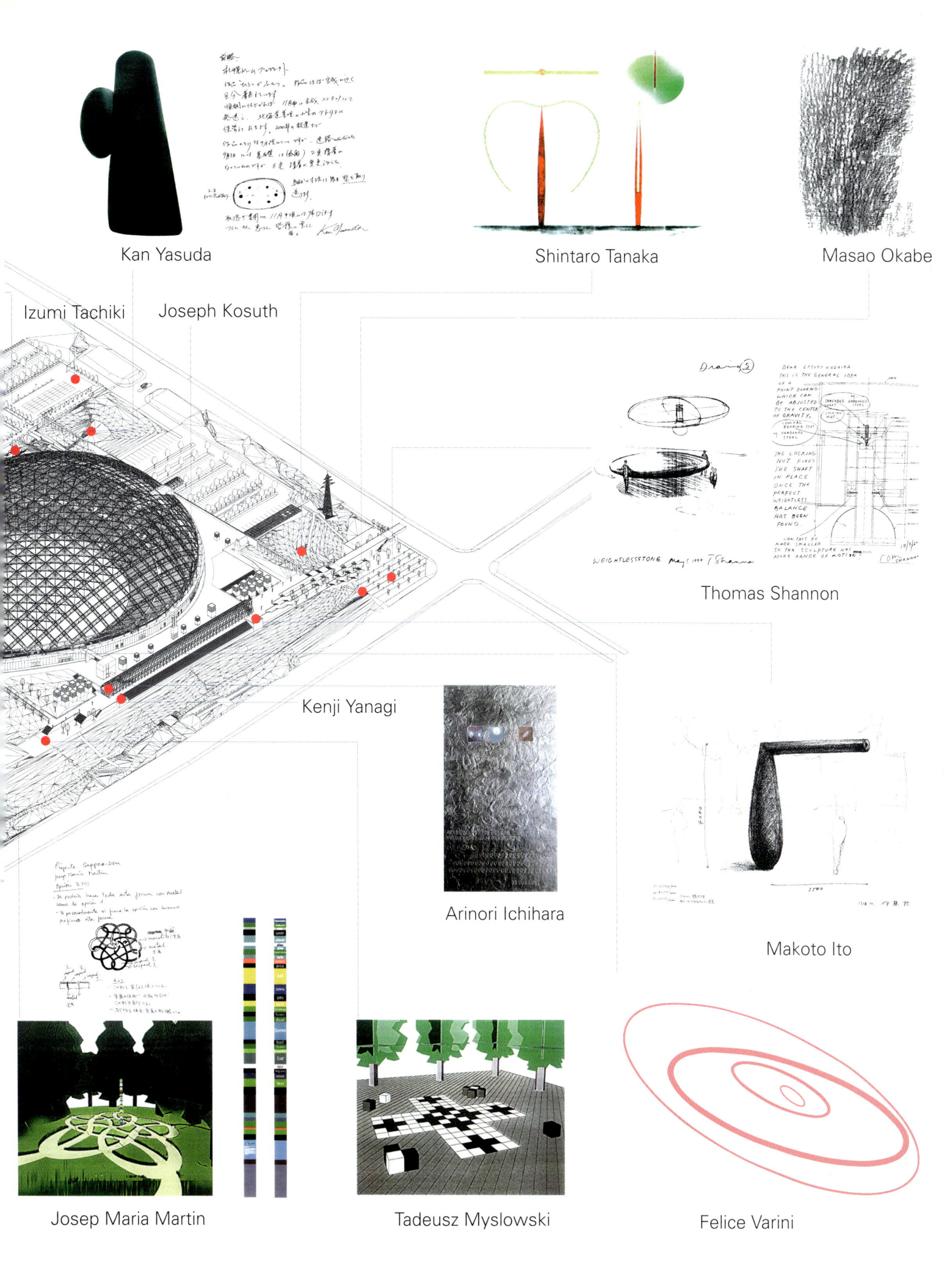

Shintaro Tanaka

Masao Okabe

Izumi Tachiki   Joseph Kosuth

Thomas Shannon

Kenji Yanagi

Arinori Ichihara

Makoto Ito

Josep Maria Martin

Tadeusz Myslowski

Felice Varini

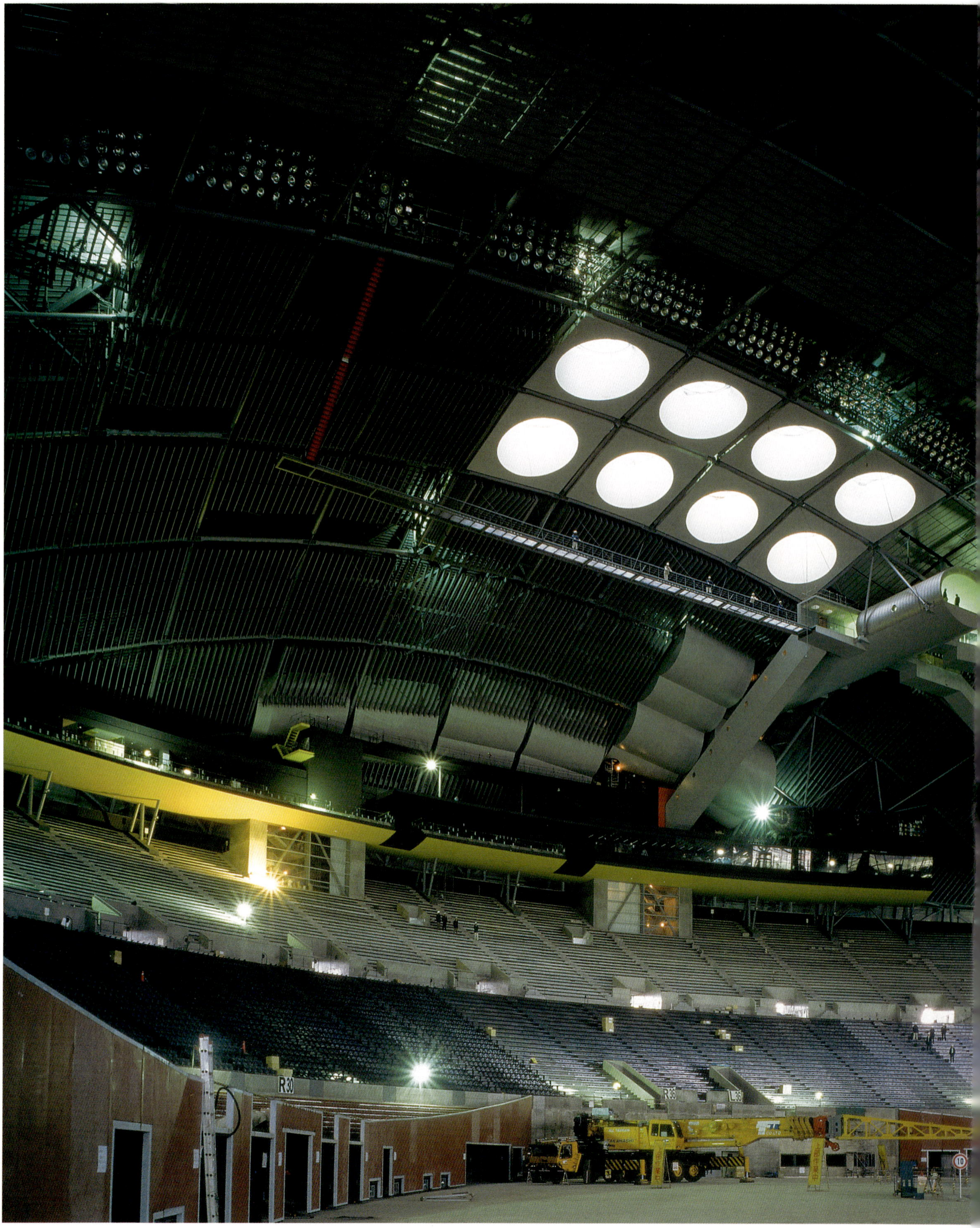

Sapporo Dome

Interior views of closed arena; upward view of escalator

Hiroshi Hara

Sapporo Dome

View towards open area across Bow Bridge from closed arena

Hiroshi Hara

Sapporo Dome

View of Bow Bridge and moving wall across open field

Hiroshi Hara

View of the concourse under the stands

Hiroshi Hara

Left: View of grass stand of open arena

Opposite: Detail of open arena and observatory

Hiroshi Hara: List of Works

The projects in bold are featured in this book

1963   Kyoto International Conference Hall, Competition, Sakyo-ku, Kyoto, 1963

1965   'The World of Yukotai' Environment, Exhibition

1967   **Ito House**, Mitaka, Tokyo, 1967

1968   Shimoshizu Primary School, Sakura, Chiba Prefecture, 1967–8; Keisho Kindergarten, Machida, Tokyo, 1967–8; Induction House (theoretical project), 1968; 'The Hole', Biennale de Paris, France, 1968

1971   Sea–folk Museum, Toba, Mie Prefecture, 1970–1

1972   **Awazu House**, Kawasaki, Kanagawa Pref. 1970–2; Fishing Cottage in Izu, Ito, Shizuoka Prefecture, 1971–2

1974   **Hara House**, Machida, Tokyo, 1973–4

1976   **Kudo Villa**, Karuizawa, Nagano Prefecture, 1974–6

1977   Kuragaki House, Setagaya, Tokyo, 1976–7

1978   **Niramu House**, Ichinomiya, Chiba Prefecture, 1976–8

1979   Shokyodo, Toyota, Aichi Prefecture, 1977–9; Akita House, Nerima-ku, Tokyo, 1978–9

1981   Sueda Art Gallery, Yufuin, Oita Prefecture, 1980–1; Mori Lithography Workshop, Sakaki, Nagano Prefecture, 1980–1; Tsurukawa Nursery School, Machida, Tokyo, 1980–1

1982   Hillport Hotel, Shibuya-ku, Tokyo, 1979–82; **Nakatsuka House** 'The Stage of Dreams', Ito, Shizuoka Prefecture, 1980–2; Parc de la Villette, International Competition, Paris, France, 1982

1983   Shibukawa Redevelopment, Shibukawa, Nagano Prefecture, 1981–3; Opéra de la Bastille (International Competition), Paris, France, 1983

1985   Shima House, Musashino, Tokyo, 1984–5

1986   **Tasaki Museum of Art**, Karuizawa, Nagano, 1983–6; **Modal Space of Consciousness**, Exhibition Installation, 1984–6; Kohagura House, Ginowan, Okinawa, 1984–96; Kitagawa House, Joetsu, Niigata Prefecture, 1985–6

1987   **Yamato International**, Ota-ku, Tokyo, 1985–7; **Kenju Park 'Forest House'** Nakaniida, Miyagi Prefecture, 1986–7; **Josei Primary School**, Naha, Okinawa, 1983–7

1988   **Iida City Museum**, Iida, Nagano Prefecture, 1986–8; **Yukian Teahouse**, Ikaho, Gumma Prefecture, 1986–8; **Media Park Köln** (Competition), Cologne, Germany, 1987–8

1989   BEEB Building, Sendai, Miyagi Prefecture, 1987–9

1990   Sotetsu Culture Center, Yokohama, Kanagawa Prefecture, 1988–90; Takezono-nishi Primary School, Tsukuba, Ibaraki Prefecture, 1988–90; Green Hall, Musashino Women's College, Hoya, Tokyo, 1988–90; **La Cité Internationale de Montreal** (Competition winner), Montreal, 1990; Palma de Majorca (Competition Entry), 1990

1991   Kindergarten, Musashino Women's College, Hoya, Tokyo, 1988–91; Friedrichstadt Passagen, (Project), Berlin, Germany, 1991

1992   Ueda Shokai Guest House 'After the Burst,' Shibuya-ku, Tokyo, 1989–92; **Ose Middle School**, Uchiko, Ehime Prefecture, 1990–2; **500m x 500m x 500m Cube** (Theoretical Project), 1992; **Future in Furniture** (Competition), 1992 (First prize); **Extra-Terrestrial Architecture** (Project), 1992

1993   **Umeda Sky Building**, Kita-ku, Osaka, 1988–93

1995   **Twenty-five Music Stands**, Exhibition Installation, 1995; Kemigawa Seminar House, University of Tokyo, Sakura, Chiba Prefecture, 1995

1997   **Miyagi Prefectural Library**, Sendai, Miyagi Prefecture, 1993–7; **JR Kyoto Station Complex**, Sakyo-ku, Kyoto, 1990–7

1998   **Hiroshima Motomachi High School**, Hiroshima, 1998; **Takagi Clinic**, Tokamachi, Niigata Pref., 1998; **Four Cube-Houses**, Machida, Nagasaki, Wakayama Prefectures, 1998

2001   **Sapporo Dome**, Sapporo, Hokkaido, 1997–2001

2002   **University of Tokyo**, Komaba Campus II, Tokyo, 1993–2002

Biography and List of Collaborators

HIROSHI HARA

1936 Born in Kawasaki, Japan
1959 University of Tokyo, BA
1961 University of Tokyo, MA
1964 University of Tokyo, PhD
      Associate Professor at Faculty of Architecture, University of Tokyo
1969 Associate Professor at Institute of Industrial Science, University of Tokyo
1970 Collaborates with Atelier Phi for design practices
1982 Professor at Institute of Industrial Science, University of Tokyo
1997 Professor Emeritus, University of Tokyo

AWARDS

1986 Annual Award of the Architectural Institute of Japan for the Tasaki Museum of Art
1987 Media Park Köln International Urban Design Competition (one of the six winners)
1987 Naha City Townscape Award for Josei Primary School
1987 AD Award for the Yamato International
1987 LUMEN Award for the Modal Space of Consciousness
1988 1st Togo Murano Award for the Yamato International
1988 Suntory Award for 'Space <From Function to Modality>' (book)
1988 BCS Award for the Yamato International
1989 The Central Districts Architectural Award for the Iida City Museum
1990 La Cité Internationale de Montréal, Montreal 1990-2000 / International Competition in Urban Design and Urban planning (one of the three winners)
1991 Design Competition for the Reconstruction of JR Kyoto Station (International Competition) (winner)
1993 Nikkei BP Technology Award Grand Prize for the Umeda Sky Building
1993 BCS Award for the Umeda Sky Building
1994 BCS Award for the Ose Middle School
1995 Annual Architectural Design Commendation of the Architectural Institute of Japan, for the Ose Middle School
1996 Sapporo Dome Design-and-Build Competition (International Competition) (grand prize winner)
1997 Annual Award for the Best Anti-Disaster Building System, for JR Kyoto Station
1998 Annual Award from the Association of Railway Building Industries, for JR Kyoto Station
1999 BCS Award for JR Kyoto Station
2000 BCS Award for Miyagi Prefectural Library

MEMBERS OF ATELIER PHI

Wakana Kitagawa, Tomoaki Ogawa

Yukio Kawase, Natsuko Higano, Jyunichi Nagamatsu, Katsuyuki Shiozaki, Masaki Harada, Tomoko Hiraoka, Kotoaki Asano, Yoshihiko Yoshihara, Yoshihito Iwasaki, Aya Yamagishi

FORMER MEMBERS OF ATELIER PHI

Jyoji Ishiyama, Masatoshi Hara, Norigazu Nagata, Yukitaka Date, Kazumi Kuruma, Keiko Ikeda, Midori Yokoyama, Norihisa Kanou, Syunji Fukumura, Kazunori Okamura, Hidemi Ihara, Yoshihito Tsuchiya, Ichiro Suzuki, Takayuki Suzuki, Daigo Ishii, Motomu Uno, Hidekuni Magaribuchi, Dai Nagasaka, Humiko Hattori, Dai Matsumoto, Soichi Shitara, Hidenobu Matsumoto, Syunji Nakatani, Takahiro Watanabe, Kunitoshihiko Tanahashi, Keitaro Hisano, Mika Takahashi, Martin van der Linden, Naoto Yaegashi, Hiroshi Yatsuo, Tsuneyuki Toyota, Yoshiaki Hitomi, Keisuke Kawamura, Amy Kimura

MEMBERS OF HARA LABORATORY AT THE INSTITUTE OF INDUSTRIAL SCIENCE, UNIVERSITY OF TOKYO

Kazuhiro Kojima, Shigeru Sakiyama, Takaaki Fujiki, Kyoko Yoshimatsu, Yasuyuki Ito, Hiroshi Horiba, Maho Hiro, Akira Kanao, Kazumi Kudo, Kiyoaki Oikawa, Masao Koizumi, Naoko Tarao, Masakuni Mitsuhashi, Weimin Huang, Atsushi Miyazaki, Hiroki Kitagawa, Momoyo Gota, Seiji Kuwabara, Motoko Takahashi, Kazuhito Furuya, Francesco Montagnana, Dai Tsukamoto, Jorge Fernandez, Kotaro Imai, Kayoko Kitae, Yuji Shimizu, Hiroshi Ota, Syunichi Shinkai, Yasumi Taketomi, Yasuhiro Minami, Manabu Okochi, Makoto Sato, Osamu Tsukihashi, Tetsuo Tsuchiya, Kenichiro Hashimoto, Shintaro Yamanaka, Kaori Ito, Masahito Ozeki, Hideo Kanatsuka, Yasukazu Kawase, Yu Tomita, Takeshi Maruyama

MEMBERS OF THE MONTEVIDEO SEMINAR (Parque Vas Ferreira)

Alejandro Sande, Eduardo Ramos, Armando Nunez, Cesar Lorenzo, Gabriel Etchepare, Mercedes Chirico, Leticia Martinez

MEMBERS OF THE WORKSHOP IN CORDOBA

Elvira Fernandez, Alejandro Soneira, Guillermo Lange, Ana Mendoza, Silvina Manzotti